AF867769

This page intentionally left blank

WWW.JOURNALOFCOSMOINTEL.COM

Interuniversal Press

The Scientific Journal of Cosmointel
Vaughan, Canada

Journal of CosmoIntel

The First Scientific Journal in T-Consciousness Research

No. 2 | Apr. 2024
ISSN 2817-6995

Publication Details

Publication Title:
The Scientific Journal of CosmoIntel

Publisher: Interuniversal Press

Intellectual Property Rights:
Mohammad Ali Taheri

Managing Director and Editor-in-Chief:
Mohammad Ali Taheri

Publication Language: Farsi - English

Year of Publication: 2024

Category: Scientific - Research

Website: WWW.JOURNALOFCOSMOINTEL.COM

Email: manager@journalofcosmointel.com

All material and intellectual rights to this special issue, including theories, concepts and foundations, terminology, innovative images, translations and other content, belong to the Scientific Journal of Cosmointel and Interuniversal Press, founded and owned by Mohammad Ali Taheri. Any scientific, literary and artistic plagiarism, republication or modification of the contents and images without written permission, and any misuse of this work, will be prosecuted by law.

All intellectual property and material rights of the issue are owned by the Scientific Journal of Cosmointel.

Interuniversal Press

Originator: Mohammad Ali Taheri

A Collection of Novel Theories in the Field of Cosmology by Mohammad Ali Taheri

Special Issue

T-Consciousness Cosmology

The Cosmic Black Hole

Important Point: A documentary concerning the topics addressed in this issue, titled:

"**The Cosmic Black Hole**"

was broadcast on

Dec 15, 2021

and is available for viewing on

the YouTube channel of Mohammad Ali Taheri

With special thanks to the research committee:

Sarah Hemmati
Mahyar Ramezanzadeh

WWW.JOURNALOFCOSMOINTEL.COM
WWW.COSMOINTEL.COM

Interuniversal Press

Preface

DOI: doi.org/10.61450/joci.v3iTC1EN.174

Introduction

Unveiling the Cosmos

The vast expanse of the cosmos, a celestial ocean teeming with constellations and swirling galaxies, has ignited an unquenchable spark of curiosity within humanity since the dawn of time. This eagerness to understand the universe in which we live has continually driven humankind to uncover the secrets of the cosmos. This endeavor is evident in ancient civilizations that charted constellations, as well as in modern astronomers who peer into the farthest reaches of space. Throughout history, pioneering astronomers like Galileo and Copernicus paved the way for a more tangible understanding of the universe. Eventually, modern cosmology, equipped with powerful telescopes and advanced detectors, has revolutionized our understanding of the universe as a whole, offering a framework for the origin, evolution, and ultimate fate of the universe.

From Philosophy to Physics Theories and Observations

The mysterious enigma of the cosmos has been and continues to be a pretense for the clash of ideas, inspiration, and countless arguments throughout human history. For tens of thousands of years, the configuration of stars in the sky has captivated our attention and inspired the creation of constellations from which myths and fictional stories were born, followed by philosophical inquiries essentially defining the philosophy of nature. Subsequently, the introduction of mathematics to simplify these philosophical concepts has opened up the world of physics to humanity. Persistent efforts and the accumulation of unanswered questions have not only compelled humans to observe and monitor the cosmos on larger scales with the advancement of technology but also aimed to interpret and analyze its components. This has been achieved through the processing of numerous mathematical models and the formulation of various theories, which sometimes align and sometimes contradict.

However, despite technical and technological advancements, the definition of the cosmos has been obscured in countless equations, calculations, models, and physics theories, with no clear and powerful unified interpretation provided thus far. Humanity remains at a loss in answering the questions: What is the cosmos? Why does it exist, and where is it headed? This is where the power of intuition and perception from a holistic viewpoint becomes essential for understanding and grasping the philosophy of the why and how of the cosmos. From this perspective, T-Consciousness Cosmology has attempted to unravel the complexities of conventional cosmology and provide a complete, clear, and unified interpretation of the cosmos through new theories. Additionally, this viewpoint addresses all concepts and the nature of what transpires in the cosmos, including Cosmic Information, Cosmic Mind, Cosmic T-Consciousness, and also various forms of Cosmic Life. These topics, however, do not entirely conform to the limited framework of currently accepted theories in physics and cosmology.

The Conventional Cosmological Perspective and Common Theories

Big Bang: The universe originated from an ultra-dense and hot point.

With the advancement of scientific research, the Big Bang theory was eventually established as the dominant cosmological model for the origin of the universe. This theory posits that the cosmos began nearly 13.8 billion years ago from an infinitesimal point with incredibly high density and temperature, known as the singularity. This singularity rapidly expanded and cooled in a fraction of a second, initiating the process of element formation and subsequently the formation of celestial bodies, ultimately leading to the cosmos we observe today.

From the perspective of cosmologists, the Big Bang theory is strongly supported by several key observations. One such observation is the discovery of the Cosmic Microwave Background (CMB). This radiation is considered a faint echo of the early universe, permeating the entire cosmos. In other words, its uniformity across the sky aligns with the predictions of a hot, dense origin of the universe. Additionally, the observed abundance of light elements such as hydrogen and helium in the universe also corresponds with the nuclear synthesis predicted to have occurred following the Big Bang.

Standard Model of Cosmology: A Framework for Cosmic Evolution

Building on the Big Bang theory, the standard model of cosmology (Lambda-CDM) provides another perspective on the universe's initial moments and its evolution, from the formation of fundamental particles to the creation of large structures like galaxies and galaxy clusters. This model incorporates the theory of inflation, a period of rapid exponential expansion believed to have occurred shortly after the Big Bang. The theory addresses the observed uniformity in the large-scale universe and theoretically resolves some of the problems with the Big Bang model. Additionally, the model relies on the existence of dark matter and dark energy, which are considered enigmatic components of the universe. Although invisible, dark matter influences the motion of galaxies and clusters through its gravitational pull. On the other hand, dark energy is believed to be responsible for the currently accelerated expansion of the universe.

Unanswered Questions and Ongoing Explorations

Despite the successes achieved in cosmology, the Big Bang theory and the standard model still face challenges. Questions about how the universe was born, the process of its evolution to its current form, the nature and geometric shape of the cosmos, its ultimate fate, the characteristics of dark matter and dark energy, and the possibility of other universes – along with countless unanswered questions – continue to drive ongoing research. Furthermore, alternative cosmological models such as the steady-state model, etc., are being examined to ensure a comprehensive understanding of the origin and evolution of the universe.

T-Consciousness Cosmology: A New Perspective on the Universe

Through its novel approach, T-Consciousness Cosmology comprises a collection of theories that examine and analyze topics such as the origin of the universe, its nature, the manner of its evolution, its fate, and hundreds of other theories.

As the name suggests, this viewpoint introduces a unique consciousness known as T-Consciousness. It posits that the universe, in addition to matter and energy, contains another element called T-Consciousness, which differs from definitions previously offered in the history of science or philosophy. From this perspective, it is argued that both matter and energy themselves arise from T-Consciousness.

Furthermore, T-Consciousness Cosmology articulates that the cosmos generally consists of two parts: frequency-based (~) and non-frequency-based (-):

The frequency-based part of the cosmos describes behavior that is periodic and non-linear, characterized by amplitude and wavelength, such that it has a non-continuous effect on the cosmos (i.e. all known types of waves and ordinary matter).

In contrast, the non-frequency-based part of the cosmos describes non-periodic and linear behavior, where the amplitude and wavelength are zero, and its effects in the cosmos are linear and continuous.

In this regard, for example, it can be said that space, gravity, and time themselves do not have a frequency effect and have a sustained effect on everything. Even if, for a moment, one of these, like gravity, were to exhibit a periodic effect, the entire cosmos would disintegrate. However, it is worth noting that the result of this influence is the emergence of particles (ordinary matter), which exhibit periodic and frequency-based behaviors. Similarly, if time itself were to have a periodic effect, the cosmos would likewise collapse in the same way, despite the fact that we have a periodic method of measurement for time (tick tock of a clock). Therefore, from this perspective, for the most part, the known physical aspect of the cosmos is periodic and frequency-based.

An important point to note is that the linearity of the impact of space, gravity, time, dark matter, and dark energy refers to the inherent influence of these factors in the universe, not the outcome of their effects.

Consequently, the frequency-based part (~) of the cosmos includes matter and energy, and the non-frequency, non-pulsing part (-) of the cosmos consists of two sections:

A- A section that in conventional cosmology is referred to with different definitions, such as spacetime, dark energy, and dark matter.

B- T-Consciousness, information, mind, life, dark life energy, etc., are parts that do not have specific definitions and are not mentioned in conventional cosmology. While from the viewpoint of T-Consciousness Cosmology, they constitute the main part of the cosmos.

Important Note: In T-Consciousness Cosmology, instead of the concept of "space-time," the term "space, gravity-time" is used, in which gravity and time are always proportionally intertwined and inseparable. In fact, the effect of gravity-time is considered as two sides of the same coin. Moreover, considering that if space did not exist, the cosmos would certainly not exist either. Therefore, from this perspective, space is considered a fundamental element of the cosmos, while it is neither matter nor energy. This means the nature of space, as one of the main components of the cosmos, is non-pulsing. This concept also applies to dark energy and dark matter, which this perspective identifies as functions of space itself.

Therefore, T-Consciousness Cosmology states that the structure of the components of the cosmos, such as dark energy and dark matter, is not composed of particles. Additionally, because of its non-pulsing nature, gravity is inherently a non-frequency element. Thus, generally, gravity is also not composed of particles (such as the hypothetical graviton particles in conventional science).

Regarding the non-pulsing nature of gravity or space, it can be noted that physics calculations show that celestial bodies with significant mass or acceleration can disturb spacetime in such a way that it appears as if gravitational waves propagate in all directions. In other words, conventional cosmology predicts that although they differ from each other, gravitational waves resulting from the spinning of neutron stars, the collision of black holes, and supernova explosions can be analyzed. However, T-Consciousness Cosmology defines what is commonly referred to as gravitational waves in physics simply as the squeezing and stretching of space due to the changing behavior of massive bodies in proximity to one another. Therefore, the changes in gravitational behavior caused by massive bodies only lead to the contraction and expansion of space. In simpler terms, gravity has a linear impact on the structure of space, not a wave-like one.

Like gravity, time exerts its influence in a linear fashion on the cosmos and its components, in tandem with gravity. If gravity were zero, time would also be zero. Conversely, if gravity approached infinity, time would similarly become infinite. It is also essential to mention that the type of timekeeping invented by humans (i.e. the ticking of a clock) is completely arbitrary, as time does not have a frequency or pulsing nature.

Consequently, T-Consciousness Cosmology uses "space, gravity-time" instead of the well-known term "spacetime."

The Origin and Fate of the Universe

Existing models in conventional cosmology have not yet provided a widely accepted theory about what existed before the Big Bang or how the various forms of matter and energy known today came into being at the initial moment of the explosion. This issue remains shrouded in ambiguity for cosmologists. In this context, T-Consciousness Cosmology, by introducing a new model named the 'Spherical Cosmos Model,' not only addresses the origin or how the initial seed of the cosmos came into existence, but also acknowledges the expansion of the cosmos and introduces a shell made of TAM (Taheri Absolute Matter), that isolates the cosmos. This model proposes a different foundational mechanism compared to conventional inflationary models and introduces a new concept called 'Space Rebound' to explain the increase in the volume of the cosmos. In fact, the theories of this viewpoint support each other in the understanding of the structure of the universe as a whole system, making simple predictions about the behavior of the cosmos. Moreover, T-Consciousness Cosmology, by addressing the nature of dark matter and dark energy and their functions, determines the cause of the cosmic expansion and its ultimate fate. Additionally, in line with the Spherical Cosmos Model (SCM), a new theory about another stage of the lifecycle of the cosmos, referred to as its 'Reversion,' is also proposed.

The Nature of the Building Blocks of the Cosmos

T-Consciousness Cosmology, in addition to addressing the general behavior of the cosmos, also explores the formation and function of its components, introducing new types of matter. According to the Spherical Cosmos Model, there is no contradiction between the formation mechanism of fundamental particles and the initial point of the cosmos (Big Bang). However, in the standard model of cosmology, which includes the theory of inflation and is based on general relativity and the standard model of fundamental particles, there is a clear contradiction known as the singularity at the birth of the universe and the formation of matter. Specifically, the singularity, a consequence of general relativity, is an obstacle that is inconsistent with the formation of fundamental particles in the initial moments of the cosmos's birth.

Overall, it can be stated that T-Consciousness Cosmology offers a unique view of the cosmos by altering the perspective of the observer. From this shifted viewpoint, the cosmos is perceived as a grand system endowed with distinct identity, personality, and behavior. This system not only follows a specific trajectory to fulfill a special purpose but also demonstrates a high level of intelligence.

The Multiverse from a New Perspective

T-Consciousness Cosmology asserts that the cosmos in which we currently live follows a sequential principle (Consecutive Cosmos) and has its own lifecycle. It is one of countless homogeneous or heterogeneous universes, each with its own unique characteristics and behaviors (laws of physics).

Additionally, from this viewpoint, the fundamental constants of physics change according to different cosmic epochs and locations. For example, gravity-time will range from infinity at the beginning of the cosmic lifecycle to zero at the terminal edge of the cosmos (the ultimate stage of space rebound).

and…

Originator of T-Consciousness Cosmology: Mohammad Ali Taheri

The Cosmic Black Hole

DOI: doi.org/10.61450/joci.v3iTC2EN.176

Abstract

Concurrent with advancements in cosmology, various theories have been proposed about the origin of the universe or its ultimate fate. These theories include the Steady State Theory, Oscillating Universe Theory or Cyclic model, Multiverse Theory, String Theory/M-Theory, Quantum Gravity or Loop Quantum Cosmology, The Big Bang Theory, Inflationary Universe Theory, and the No-Boundary Proposal. Among these, the Big Bang theory has been widely accepted by most scientists, with the Inflation theory serving as its complementary addition. On the other hand, theories such as the big rip, flat universe, and big crunch have also addressed the ultimate fate of the cosmos. However, T-Consciousness Cosmology presents a new hypothesis to explain how the universe was born from a black hole named the *'Cosmic Black Hole,'* or the initial seed of the universe. This hypothesis not only addresses how this particular type of black hole forms, contingent on the reversion of the cosmos according to the Spherical Cosmos model but also details its fundamental differences from known black holes, referred to as *"intra-cosmic black holes."* The reversion of the cosmos in the 'Spherical Cosmos Model,' is described through a mechanism known as space Rebound, which is distinct from the Big Crunch. The Terminal Edge of the cosmos is defined as the maximum radius at which space mesh is capable of rebound during the universe's volume increase. In the process of the rebound of space to its ultimate extent, objects within the cosmos face complete disintegration and transform into waves called absolute waves. Due to the inherent rotation of the cosmos and its reversion from the Terminal Edge, these waves collide and create gravitational centers in the central regions of the spherical cosmos, forming new types of matter known as light-dark matter, dark-dark matter, and thermal matter. Each of these mentioned materials represents a new type of matter introduced by T-Consciousness Cosmology. Additionally, according to this model, the cosmos contracts into a very tiny point with a final quench, where all types of matter and fundamental forces unite, forming a new type of absolute matter called *'Taheri Absolute Matter'* (TAM). The spherical cosmos model also divides time into various types: 'Longitudinal' and 'Transverse,' and unlike the theory of relativity, where time is considered a dimension, it classifies it as one of the types of transverse time introduced as an entropic force. In essence, this type of force (time) acts against the force of gravity and, by disintegrating all types of objects from fundamental to large scale, acts as an agent of stress or tension release from the space mesh.

Keywords: Cosmic Black Hole, Spherical Cosmos Model, Cosmic Rebound, Terminal Edge of the Cosmos, Absolute Waves, Intrinsic Rotation of the Cosmos, Light-Dark Matter, Dark-Dark Matter, New Thermal Matter, Absolute Matter or (TAM), Longitudinal Time, Transverse Time, Entropic Time Force, Space Stress or Tension, Final Cosmic Quench

Theories in Conventional Cosmology on the Beginning of the Universe

Introduction

The origin of the universe or how it came into existence, as well as its ultimate fate, has always been one of the most intriguing and challenging questions in science and philosophy. To address this challenge, it can be said that the basis for many hypotheses in the past was derived from discovery and intuition or philosophy, with various theories being proposed in this regard. It is important to note that the validity of each of these theories must be demonstrated with observational evidence.

Other theories, such as the big rip, big crunch, and flat universe, also propose possibilities that the cosmos may face in the future. In contrast, the Big Bang theory describes the beginning of the universe, which has been continuously refined throughout history and, despite its flaws, has so far been agreed upon by the majority of cosmologists. In cosmology, several theories besides the Big Bang have been proposed about what happened before the Big Bang or how the universe began, none of which have been definitively proven or widely accepted to date. Among these theories are:

1- **Cosmic Inflation:** This theory suggests that the universe, in its first moments when it was in a state of unstable energy, underwent a very rapid expansion. In other words, the universe increased in volume by several times in a fraction of a second, due to a mysterious type of energy that permeates empty space. The inflation theory predicts some obscure issues such as the uniform distribution of energy and the flat geometry of space-time. In fact, this theory states that inflation smoothed out the initial irregularities and laid the foundation for the cosmic structures that we observe today. Inflation ended with an explosion of radiations, which are today detected as the Cosmic Microwave Background radiation (Figure 1).[1 2]

Figure 1: Inflation theory in the Big Bang Model

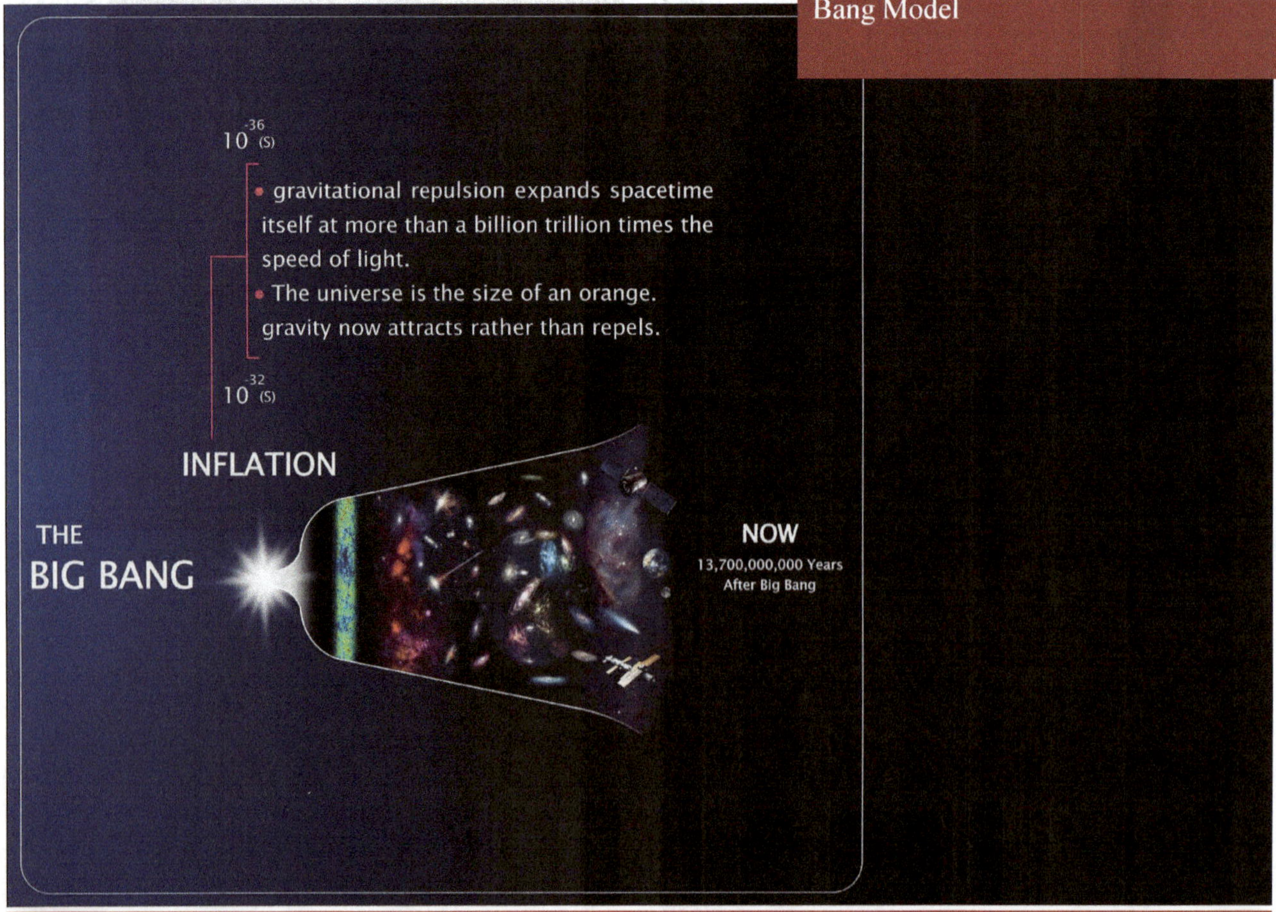

2- **Quantum Fluctuations:** Quantum fluctuations are based on the principles of quantum mechanics. This theory suggests that the universe could have originated from quantum fluctuations in the vacuum of empty space, often referred to as quantum nothingness. In essence, fluctuations can cause the creation of particle and antiparticle pairs, which have the potential to lead to the creation of the entire universe. According to this theory, in the early cosmos, quantum fluctuations transformed infinitesimally small changes into significant differences in the distribution of matter on cosmic scales during the inflationary period. These changes later served as the foundation for the structure of the universe. Essentially, quantum fluctuations are constantly occurring, small and random variations in energy and matter fields. One of these fluctuations could have started the Big Bang and created a bubble of space-time that expanded into our universe (Figure 2).[3, 4, 5]

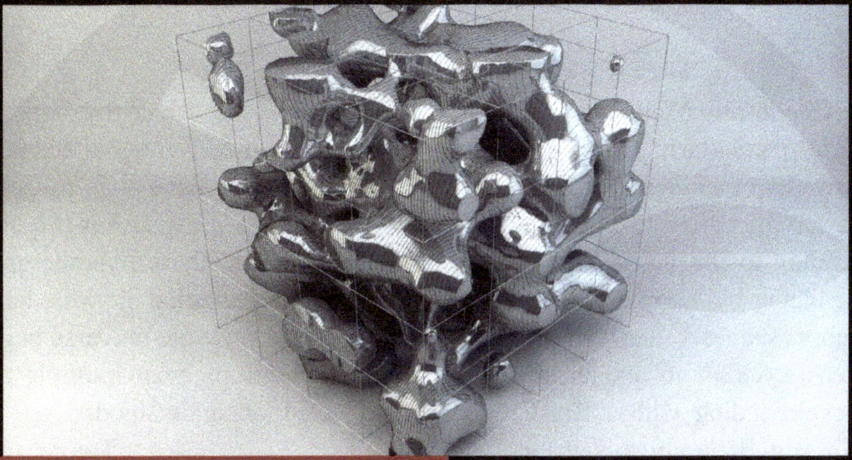

Figure 2: A three-dimensional rendition of quantum fluctuations (or spacetime foam) at subatomic scales Augmented structures, 10-33 cm passage, Credits: Refik Anadol and Alper Derinbogaz

3- **No-Boundary Proposal:** This theory, presented by Stephen Hawking and James Hartle was first proposed in 1983, when it was later developed and completed, is about the state of the universe before the Planck era and suggests that the universe did not begin at a singular point in time and has no initial boundary. It is based on the idea that if we go back to the initial moment of the universe, the concept of time becomes obsolete. Therefore, the No-Boundary Proposal illustrates that measuring events before the Big Bang is impossible. Scientists employ a mathematical tool called imaginary time to describe the universe as a smooth, four-dimensional surface without any singularities or edges. In this view, there is no clear distinction between before and after the Big Bang. Imaginary time is a mathematical concept that appears in some theories of physics, such as quantum mechanics and general relativity. Essentially, this type of time allows us to imagine that there was no Big Bang and that the beginning of the universe occurs as a slow transition from one state to another (Figure 3).[6, 7, 8, 9]

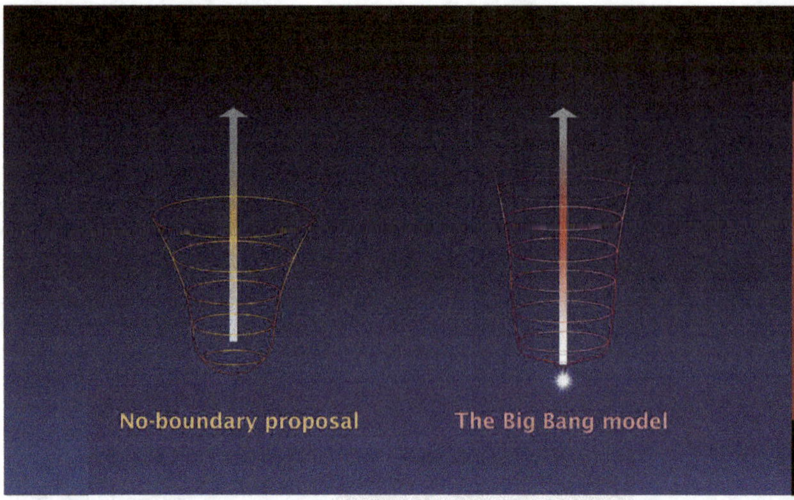

Figure 3: [Left image] A borderless universe with a one-dimensional spatial section. The no-boundary proposal states that one should consider geometries in which there is no boundary, and which have been smoothly "rounded off" in the past. [Right image] The beginning of the universe from the point of singularity according to the Big Bang theory.

4- **Cyclic Model or Oscillating Model:** In the 1920s, theoretical physicists, particularly Albert Einstein, considered the possibility of an eternal cyclic model for the universe as an alternative to the expanding universe model, leading to the introduction of the oscillating universe theory by Alexander Friedmann in 1922. This theory suggests that the universe undergoes repeated cycles of expansion and contraction, each cycle ending with a Big Crunch and starting with a Big Bang. In this theory, the cycles could be driven by dark energy. From the perspective of conventional cosmology, dark energy is an unknown force that causes the acceleration of the universe's expansion (Figure 4).[10] It's worth mentioning that these hypotheses are just some of the possible scenarios for what happened before the Big Bang or how the universe began. There may be other theories or even multiple realities yet to be discovered or understood.

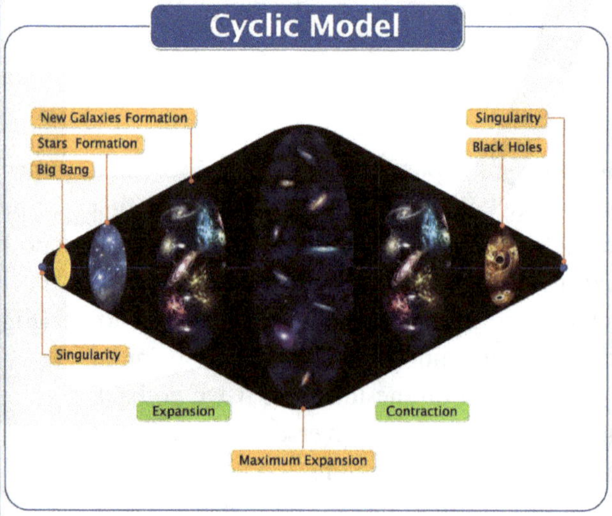

Figure 4: Rendition of the cyclic or oscillating universe model

Additionally, several other theories such as the Multiverse Theory[62, 63], Steady State Theory[64], String Theory[65] or M-Theory, and Loop Quantum Cosmology Theory[66] have been proposed, each offering a different perspective on the manner in which the universe came into being.

The Cosmic Black Hole Hypothesis or the Initial Seed of the Cosmos, a New Proposal in Cosmology

T-Consciousness Cosmology, by examining the mechanism of the cosmos from a different perspective, offers a new analysis of the genesis, expansion, and fate of the cosmos by presenting a spherical cosmos model and new hypotheses.

According to this model, billions of years ago, the universe was born from a black hole named the Cosmic Black Hole; meaning that all types of matter, energy, or fundamental forces that we know, before the existence and formation of the structure of the current cosmos, were in complete unity in a very small point, forming an 'absolute matter' which is the content of the Cosmic Black Hole. This black hole has a completely different structure and definition, unimaginably small and significantly different from the known black holes within the cosmos today. In other words, T-Consciousness Cosmology, by introducing the spherical cosmos model, states that this type of black hole is not only the beginning of the birth of the cosmos but also the expansion and formation of various types of matter and energy leading to the creation of countless cosmic bodies have started from and can be defined within this black hole. Therefore, according to the spherical cosmos model, the Cosmic Black Hole, or the initial point of the cosmos, has major differences from the known black holes. To express these differences, we first delve into defining and examining the intra-cosmic black holes presented in conventional science.

Figure 5: Artistic rendition of a Black Hole

Definition of a Black Hole:

From the perspective of conventional cosmology, black holes are the natural outcome of general relativity and serve as a powerful analytical tool to investigate both macroscopic and microscopic properties of the universe. In essence, black holes are objects whose gravitational field is so strong that not even light can escape. These objects first attracted attention in the 18th century through John Michell and Pierre-Simon Laplace, and ultimately, Karl Schwarzschild was able to describe the features of a black hole in 1916, using Albert Einstein's general relativity.[11, 12]

In the basic definition of black holes, it can be stated that there are regions of space-time where the gravitational potential exceeds the square of the speed of light. In other words, according to general relativity, if a mass is sufficiently compressed, it can cause space-time to curve and deform, leading to the formation of a black hole. The event horizon of a black hole is its outer spherical boundary, often considered its surface, and is the region around a black hole where the gravitational effect becomes so strong that not even light can escape.[13, 14]

The adjective "black" in black holes comes from the fact that all light passing the event horizon is trapped. Therefore, a black hole behaves similarly to a black body in thermodynamics.[15, 16]

Due to this behavior, the identification of a black hole often relies on observing the motion of objects around it. If there are celestial bodies such as stars or planets in the vicinity of a black hole's trajectory, then all of these objects, if sufficiently close to the black hole, will be drawn into it. Alternatively, images captured of these objects apparently located behind the black hole will be accompanied by distortion (Figure 5).

Black holes are often formed from the death of massive stars. In fact, when such stars reach the end of their lives, they lead to colossal explosions. The mechanism of these explosions is such that due to very high heat, in addition to hydrogen and helium, carbon, oxygen, and silicon can also burn as nuclear fuel inside these stars, ultimately leading to the production of iron through continuous fusion. Iron is the most stable nucleus among the elements and does not easily participate in fusion. In fact, this element is the last atom that the core of a very hot star can produce, signifying the end of the nuclear fusion process. The lack of fuel for fusion leads to a decrease in the star's temperature, which increases the speed of the star's collapse due to gravity, eventually leading to the star's complete collapse onto itself. Finally, in an explosion known as a supernova, the star disperses its material into the surrounding space, leaving behind a remnant that can be a neutron star, pulsar, or magnetar. However, if the remaining mass of the star is several times greater than the mass of the Sun, the star will begin to collapse and during this process, the pressure of the neutrons will not be sufficient to stop it, and instead of forming a neutron star, the star's core proceeds to form a gravitational singularity. The gravity of this singularity is so powerful that it overcomes all other forces to the extent that it creates a field in space-time from which not even light can escape. This is why it is called a "black hole" (Figure 6). Relatively small black holes can also be created through the merger of neutron stars. Even a neutron star can merge with a black hole to create a larger black hole, or two black holes can collide and merge to form a more massive black hole.[17, 18]

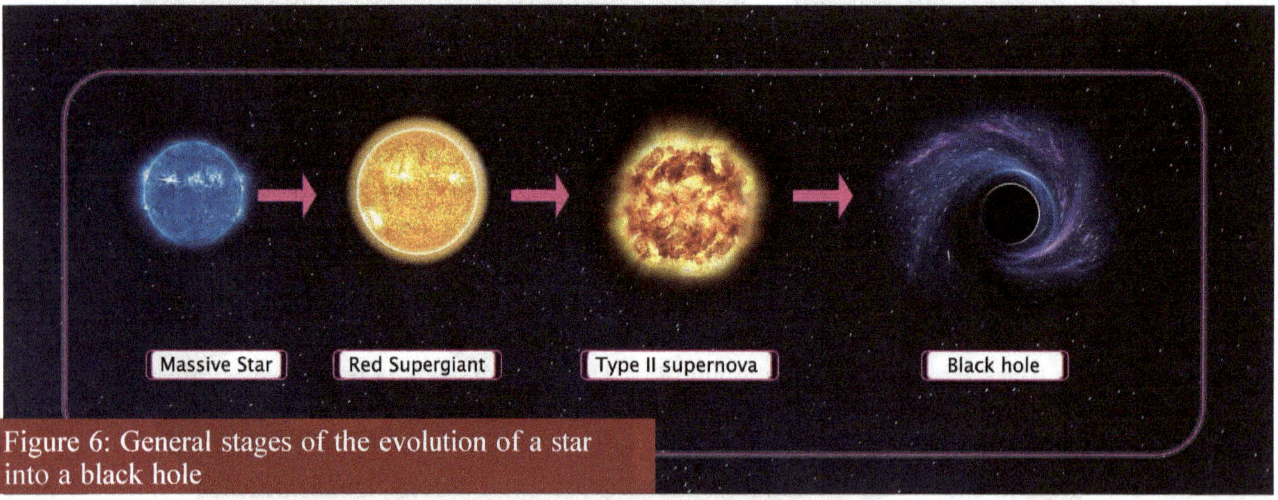

Figure 6: General stages of the evolution of a star into a black hole

Structure of a Black Hole:

Black holes have structures that can be divided into different sections:

- **Quiet Region or Static Limit (Negligible gravitational influence):** The static limit is an area around the event horizon of a rotating black hole, such as the Kerr black hole, where any trapped object will have a spin. In other words, within the static boundary, an object is forced to rotate alongside the black hole like a whirlpool. The static limit touches the event horizon at the poles of the black hole, where there is no rotational force. It is the outer boundary of the ergosphere, which, in a non-rotating black hole, aligns with the event horizon.[19, 20, 21]

- **Ergosphere:** The ergosphere is located after the quiet region. This area is an energy-rich region where light in that area both orbits around the hole and is pulled towards its center by gravity.[22]

- **Event Horizon:** The event horizon is a boundary around a black hole beyond which no light or other radiation can escape. It is essentially a point of no return.[23, 13]

- **Gravitational Space-Time Distortion:** This distortion is the geometric bending of space-time in a field created by a very dense object.[24,25,26]

- **Singularity:** A point at the center of a black hole where mass is infinitely compressed, and space-time and known laws of physics completely lose their nature (Figure 7).[27]

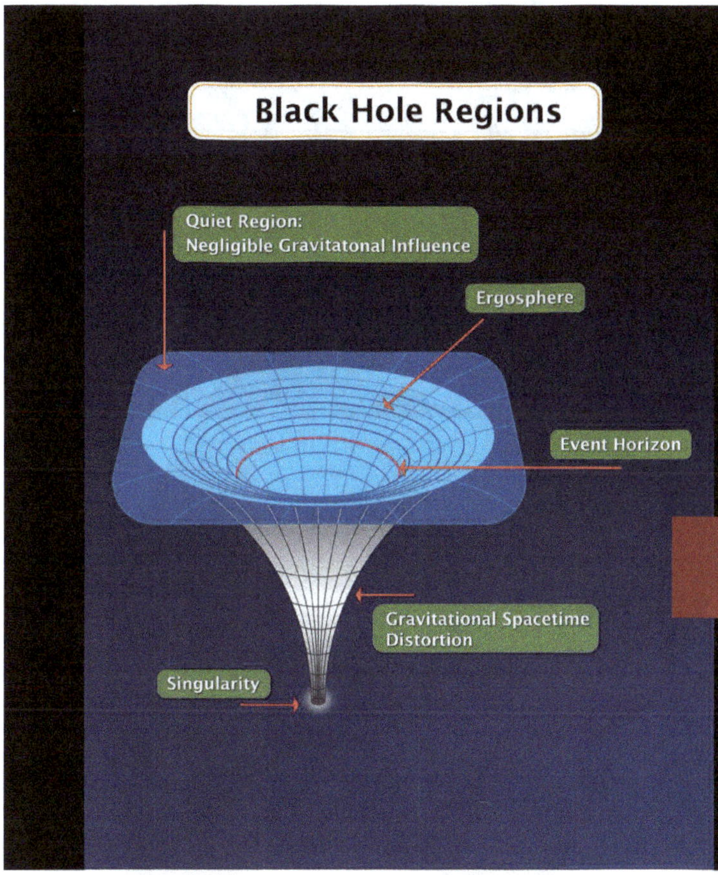

Figure 7: General Structure of a Black Hole

Also, the anatomy of a black hole from the Schwarzschild perspective includes the following elements:

- **Counter Rotating photon sphere and corotating photon sphere:** Both are regions where light is trapped by the black hole's gravity and begins to orbit it spherically.

- **Outer Event Horizon**

- **Inner Event Horizon (Cauchy Horizon)**

- **Singularity**

- **Schwarzschild Radius:** The distance between the outer event horizon and the singularity point is referred to as the Schwarzschild radius. [28,29,30]

Black holes can also be classified according to their different sizes and behaviors:

Schwarzschild Black Hole: Plays a significant role in Einstein's gravity and is the simplest type of a black hole with no charge or spin, but has an event horizon and a photon sphere. This black hole includes a singularity, meaning a point where matter is compressed up to infinite density (Figure 8).[29]

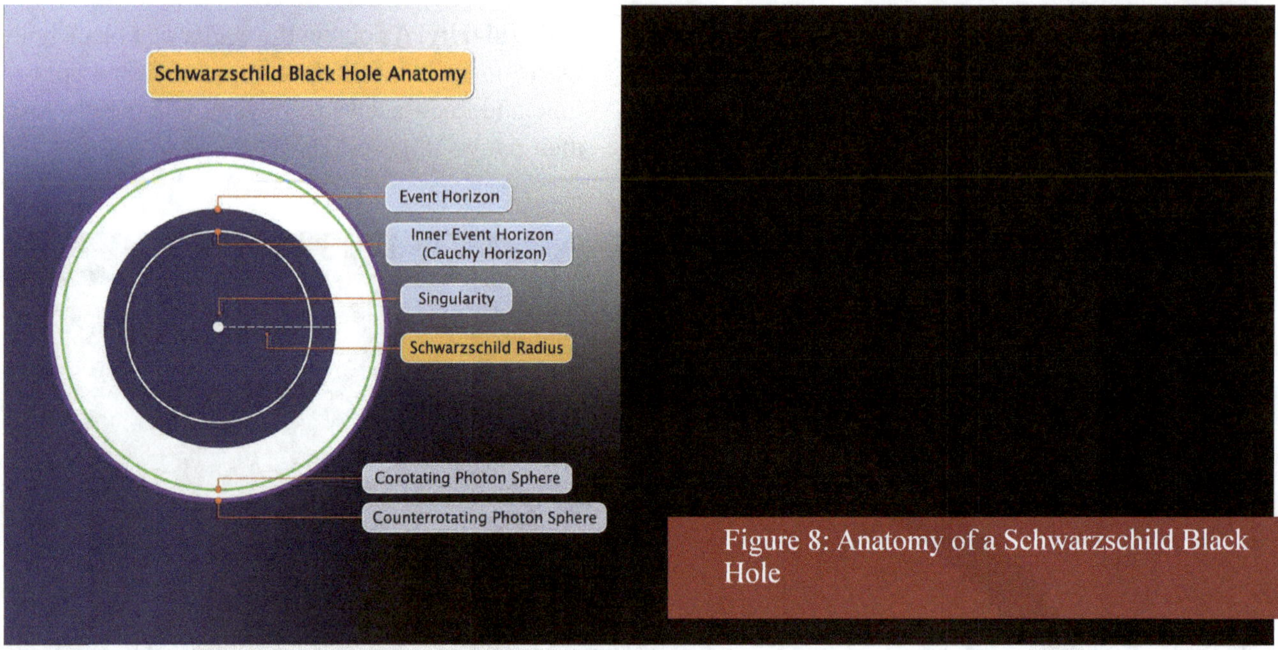

Figure 8: Anatomy of a Schwarzschild Black Hole

- **Reissner-Nordström Black Hole:** This type of black hole has both charge and spin. It also has two event horizons and one photon sphere. Additionally, it includes a point singularity, which is an improbable occurrence since its charges would neutralize each other (Figure 9).[31]

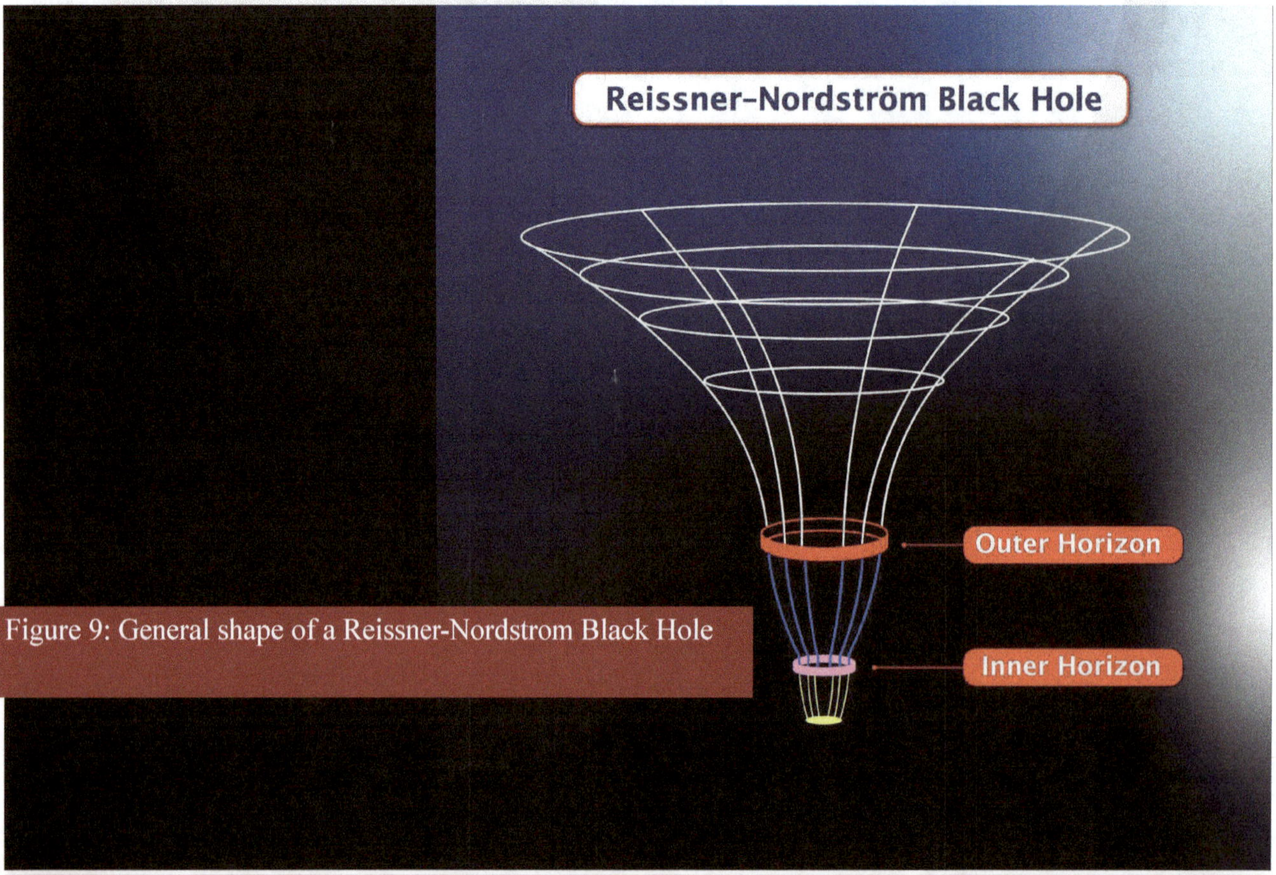

Figure 9: General shape of a Reissner-Nordstrom Black Hole

- **Kerr Black Hole:** A rotating black hole without charge, having an ellipsoidal shape. The dark area between the event horizon and the static limit, known as the ergosphere, can be used to extract energy. This type of black hole can have two event horizons and two static limits and ultimately two photon spheres that include a ring singularity. [32,33,34]

- **Kerr-Newman Black Hole:** Has both charge and spin. This type of black hole is a variant of the Kerr black hole, except it has a charge. Its structure is similar to that of a Kerr black hole, from which energy can be extracted; it also has a ring singularity (Figure 10).[35]

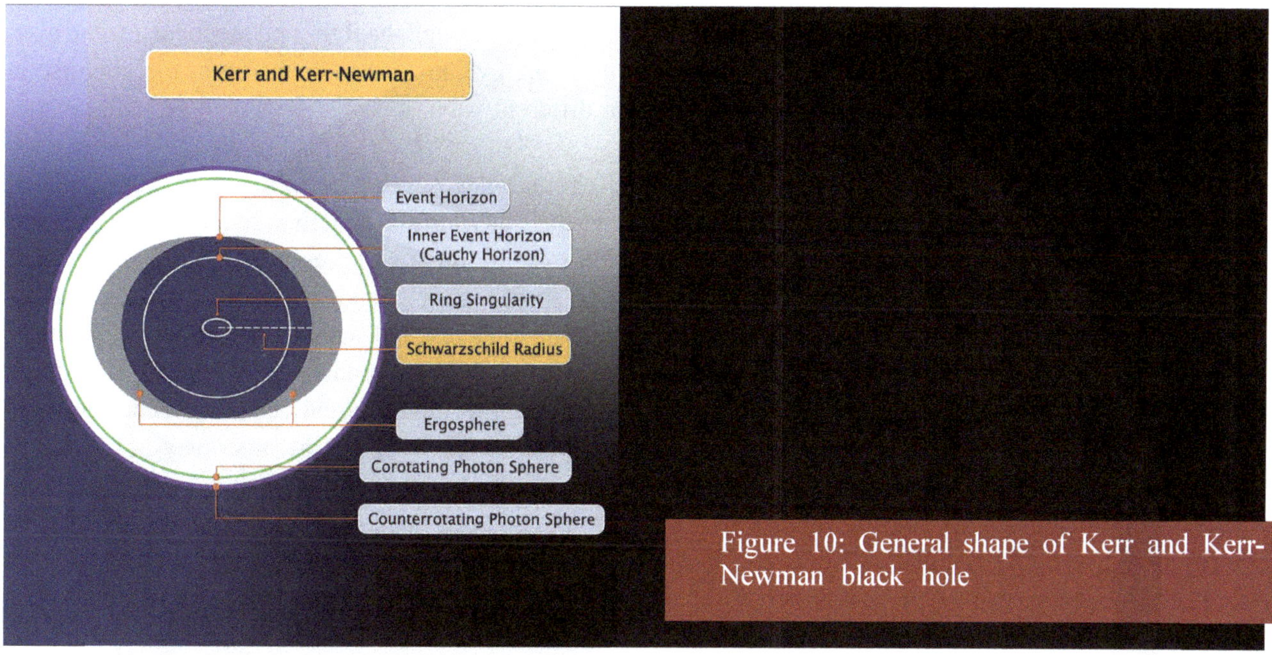

Figure 10: General shape of Kerr and Kerr-Newman black hole

In general, black holes are not directly visible to the human eye, and one method of identifying them in space is through the gravitational lensing effect, which distorts the light from stars passing near these celestial objects. Additional methods for identifying black holes include the detection of X-rays and gamma rays emitted when a star collapses into a black hole, or analyzing the oscillatory motion of stars with a black hole as an invisible companion, to pinpoint the location of these massive objects in space.[36,37,38,39]

Cosmic Black Hole or the Initial Seed of the Cosmos (T-Consciousness Cosmology)

Despite some opposition, the Big Bang theory or the Standard Cosmological Model has been accepted by most scientists in describing the mechanism of cosmic expansion, how various types of matter and energy are formed, and the formation of large structures in the universe. T-Consciousness Cosmology, however, challenges this theory and introduces a new model called the 'Spherical Cosmos.' The characteristics of this model are defined by the following hypotheses such as the Cosmic Black Hole, the Shell of the Cosmos, Space-Gravity Time, Cosmic Rotation, Center of the Cosmos, etc.

The Cosmic Black Hole hypothesis, or the initial seed of the cosmos, has an anatomy or structure very different from the known black holes. The Cosmic Black Hole hypothesis uses the Spherical Cosmos Model to address questions about the origins of the universe that remain unanswered by the Big Bang model. According to this hypothesis, the cosmos does not follow the fate of a flat universe or a big rip but has a return mechanism that is completely different from the fate of the big crunch, which considers gravity as the factor of the cosmos's re-contraction.

Formation of the Cosmic Black Hole

In conventional cosmology, the increase in the volume of the universe is considered to be an internal expansion, characterized by the increase in the metric distance between objects, which is powered by the negative pressure of dark energy. However, from the perspective of T-Consciousness Cosmology, the universe is undergoing a process called 'Rebound.' In fact, the agent of the increase in metric distance is explained through the Rebound mechanism, and also, dark energy causes the expansion of the universe in a different way.

Rebound means returning to the natural and original state of anything. To define rebound, we can use the example of a compressed spring. If we release the spring slowly and continuously, it gradually returns to its original state: free of tension and stress. This process is referred to as the spring's rebound. According to the spherical cosmos model, the same happens to the cosmos. In other words, space, as the primary bedrock of the cosmos, is transitioning from a state of compression or contraction, moving towards a state of rest, free from any tension or stress caused by gravity in space mesh (Figure 11).

Figure 11: The rebound process of the cosmos in the Spherical Cosmos model

According to the spherical cosmos model, after the complete Rebound of space, the cosmos will have a terminal edge or final boundary, which T-Consciousness Cosmology calls the "Terminal Edge of the Cosmos." In this model, it is proposed that as the volume of the cosmos increases, galaxies, stars, and celestial bodies in the direction of space rebound, toward the terminal edge of the spherical cosmos, and their speed gradually increases until they reach or even surpass the speed of light. It is evident that before reaching the speed of light, these objects will disintegrate and decompose into waves. As this process continues, the waves at the terminal edge of the cosmos become "Absolute Waves," which have wavelengths and amplitudes close to the expanse of the cosmos and rotate in alignment with the "Intrinsic Rotation of the Cosmos" (Figure 12). The cosmos demonstrates an inherent quality of omni-axial spin and rotation. This attribute is reflected in the behavior of its constituent elements. Moreover, the spin and rotation of these elements exhibit complete mutual influence. The topic of cosmic rotation will be examined in another discussion.

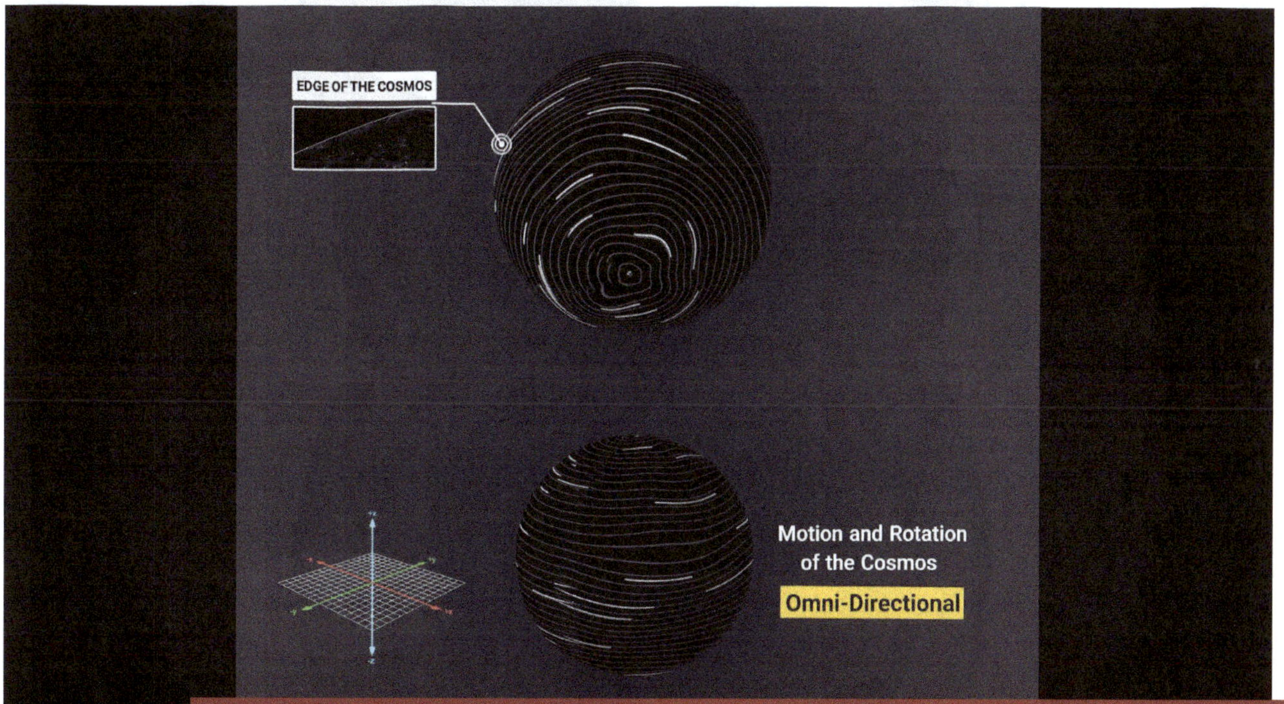

Figure 12: (Top image) The Terminal Edge of the Cosmos after the complete rebound of space mesh. (Bottom image) The rotation of the cosmos in the spherical cosmic model.

During the rebound process, absolute waves become completely parallel and tangent to the terminal edge of the spherical cosmos, signifying the point of maximum space mesh expansion. Driven by the cosmos' intrinsic rotation, these absolute waves then reverse direction from the edge, returning towards the interior over billions of years. Along this inward journey, absolute waves eventually collide with each other, within a radius of the cosmic sphere. These collisions lead to the formation of mass (ordinary matter) and consequently, gravity. The presence of this mass and gravity causes stress and tension on the space mesh, which had previously reached a natural, decompressed state at the final moment of the rebound process. The effect of this tension propagates back towards the terminal edge of the cosmos.

According to T-Consciousness Cosmology, space in the cosmos consists of infinite elastic, invisible, and dark mesh capable of rotation and twisting across vast dimensions. Therefore, the stress resulting from these collisions can be transferred across the space mesh to the very end of the terminal edge of the cosmos (Figure 13).

Figure 13: The rebound process of the cosmos in the Spherical Cosmos model

As a result of the successive collisions of waves within the space mesh occurring throughout the cosmos, dense waves emerge, giving rise to numerous gravitational centers. In other words, T-Consciousness Cosmology attributes the formation of mass to the formation of dense waves through the collision of non-dense waves. These centers are attracted to each other, creating more powerful gravitational centers that lead to the formation of light-matter (ordinary matter), and then, with the continuation of the cosmic reversion process, result in the creation of light-dark matter, which is the highly compressed ordinary matter. This event causes the space mesh to be pulled toward the gravitational centers. The process of compression or contraction of space not only creates dark matter but the continuation of this compression leads to the formation of another type of matter known as dark-dark matter.

Therefore, T-Consciousness Cosmology introduces two types of dark matter:

1- Light-Dark Matter: This is the result of severe contraction of waves or known ordinary matter. It is so dense that no light passes through it, and the reflection of light is also impossible. This type of matter is found in the Cosmic Black Hole, the shell of the cosmos, and in the central regions of intra-cosmic black holes.

2- Dark-Dark Matter: This results from the contraction of space itself, initially forming dark matter. When space is compressed again with great intensity, it forms dark-dark matter. This type of matter is also found in the Cosmic Black Hole, the shell of the cosmos, and in the central regions of intra-cosmic black holes, which will be further discussed in the hypothesis of the cosmic shell.

During the reversion of the cosmos, ordinary or light matter, which is the result of the collision of non-compressed waves with each other and is formed at gravitational centers, will have a visible frequency over a brief period of time. After that, with the decreasing volume of the universe, due to compression and conversion to light-dark matter, it will no longer be visible. Simultaneously, dark-dark matter formed from the extreme compression of space, moves toward the gravitational centers from all sides. The merging of this type of matter intensifies the gravitational strength of these centers. These centers merge as the size of the cosmos decreases, forming a unified center, which eventually leads both dark-dark matter and light-dark matter to accelerate toward this unified gravitational center, heading toward a cataclysmic collision or a tremendous quench. As a result, space is also compressed under the influence of these gravities.

As a result of increasing gravity, black holes form in various parts of the cosmos. These black holes continuously merge with each other, and instead of getting bigger, they become smaller and smaller due to the decreasing volume of the universe. Under these conditions, on the one hand, gravity also affects temperature by preventing the emission of *'Thermal Waves,'* while the pressure resulting from the contraction of space compresses these waves intensely. This is where a rare phenomenon occurs; thermal radiations become unimaginably dense, change in nature, and a new unknown type of matter emerges. During this process, space, which tends to contract due to the creation of gravitational force, and is moving towards a single center of gravity, undergoes a massive compression, and all components of the cosmos are crushed towards that gravitational center.

The weak nuclear force and the strong nuclear force disappear with the disappearance of fundamental particles. In these conditions, the concepts of wave, particle, and electromagnetic field will cease to exist. In other words, all particles and waves lose their nature, and the Planck length will also become meaningless in this unimaginably intense compression. Therefore, inside the Cosmic Black Hole, an absolute matter is formed by the extremely high compression, from the unity of materials such as a new type of matter known by T-Consciousness Cosmology as *'Thermal Matter'* (highly compressed thermal waves), light-dark matter, and dark-dark matter. In other words, the matter forming this Cosmic Black Hole is indivisible. T-Consciousness Cosmology refers to this absolute matter as *'TAM (Taheri Absolute Matter).'* Hence, TAM results from the unimaginable compression of waves and space mesh; it is the matter that makes up the infinitely small Cosmic Black Hole.

Concurrently, a specific type of time, which T-Consciousness Cosmology introduces as an "entropic force" acting to release space from stress and compression, comes into play. This type of time is introduced as the fourth dimension in the theory of relativity, but T-Consciousness Cosmology considers it a "force" and not a dimension, as it has both magnitude and direction. In the Cosmic Black Hole, time is at its infinite and maximum state to release gravity from stress. However, since there will be no mass by the end of the cosmic rebound and gravity reaches zero, and the space mesh themselves will not have any contraction, this same force of time will also reach zero. More details about the force of time will be provided in the 'Space-Gravity Time' hypothesis.

In general, T-Consciousness Cosmology states that the Cosmic Black Hole hypothesis is the infinitesimally small initial seed of the cosmos before its birth, which not only has unimaginably high gravity, but is also made of absolute matter or TAM, which is potentially ready to create all components of the universe. In other words, the massive compression of the reversion of the cosmos, in which all kinds of matter and energy lose their nature, the Cosmic Black Hole or the *'Black Pearl of the Cosmos,'* which is the beginning of a new *'Consecutive Universe,'* is born (Figure 14).

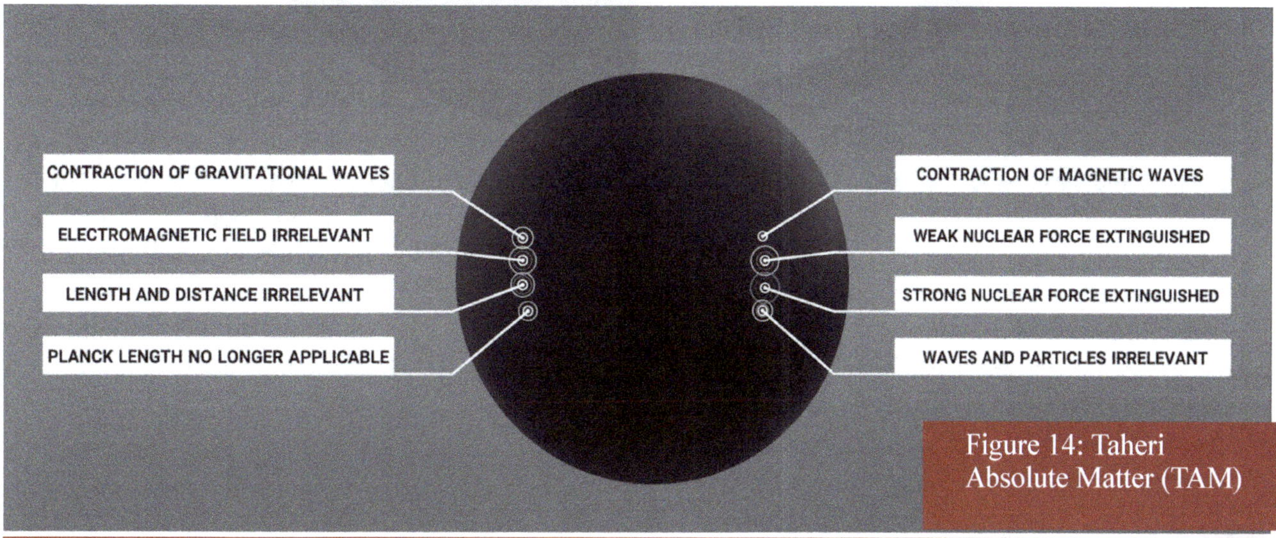

Figure 14: Taheri Absolute Matter (TAM)

On the other hand, this perspective states that the total reversion or contraction of the cosmos equals $t_{total}=t_1+\Delta t_2$. With this assumption, the cosmos, which reaches its ultimate size in n billion years, will take n billion years plus Δt_2 to form the Cosmic Black Hole, which is the total reversion time of the cosmos and its conversion into this type of black hole. The time Δt_2 is an additional period applied to the entire cosmic system due to very high gravity in moments close to the formation of the Cosmic Black Hole.

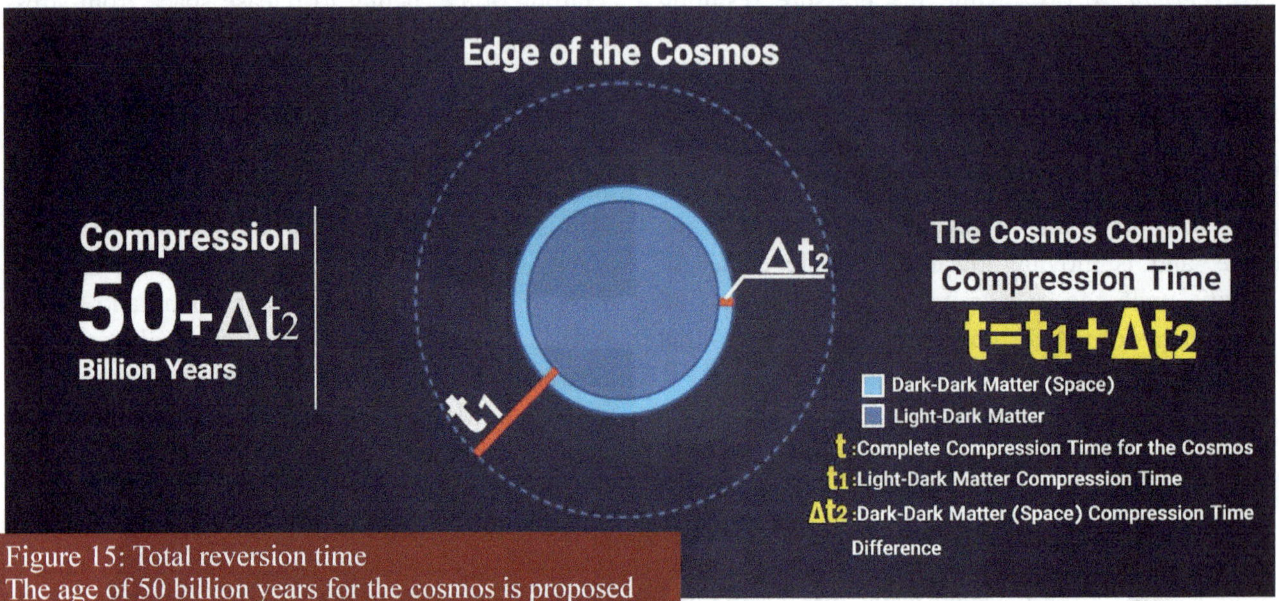

Figure 15: Total reversion time
The age of 50 billion years for the cosmos is proposed as an example in this model.

An important point here is that, from the perspective of T-Consciousness Cosmology, space and time do not exist outside this black hole, and the entire cosmos is compressed within it. In fact, the Cosmic Black Hole is so small, it is undefinable by any of the measurements known to us (i.e., Planck length).

Comparisons between the Cosmic Black Hole and Intra-Cosmic Black Holes

- **Comparison from an Absolute and Relative Perspective**

Intra-cosmic black holes have relative and variable parameters and have since existed in various types and countless numbers within the cosmos.

T-Consciousness Cosmology posits that the inevitable principle of motion ensures nothing in the cosmos remains constant. Primary and secondary constituents like space, gravity, time, matter, energy, and derivatives like temperature and light are in a state of continual change and are relative. This relativity applies to characteristics within intra-cosmic black holes. However, the Cosmic Black Hole, or the initial seed of the universe is an exception. Upon the reversion of the cosmos after its ultimate rebound, this black hole will absorb the entire cosmos and everything it encompasses is in a state of unity, homogeneity and indivisibility. Because it encompasses all the constituents of the material universe that exist in a potential state within it and has no exchange with anything external to itself, this black hole is consequently considered an absolute black hole, unparalleled and incomparable to anything else (Figure 16).

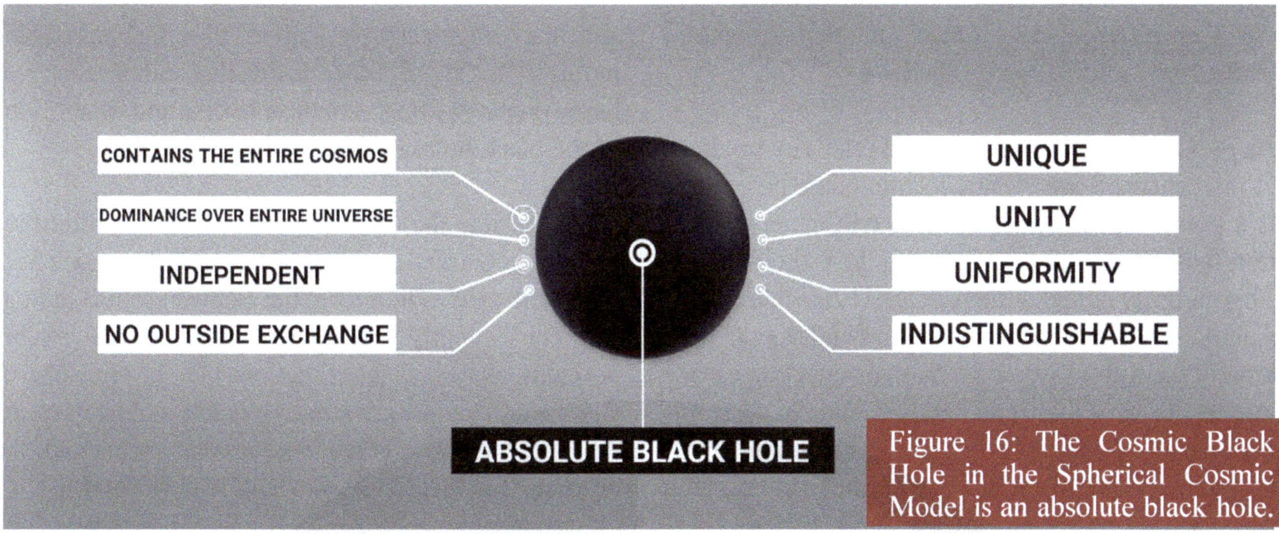

Figure 16: The Cosmic Black Hole in the Spherical Cosmic Model is an absolute black hole.

- **Comparison in terms of Planck Length, Pauli Exclusion Principle, and Chandrasekhar Limit**

The Pauli Exclusion Principle and the Chandrasekhar Limit can be employed to define and calculate the formation of intra-cosmic black holes. The Pauli Exclusion Principle states that two or more identical fermions (particles with half-integer spin) cannot simultaneously occupy the same quantum state within a quantum system. [40] The Chandrasekhar Limit is responsible for the electron degeneracy pressure that can halt the collapse of a star, resulting in the formation of a white dwarf, or a neutron star, or a black hole. In essence, it implies that two electrons can never be in the same place at the same time. It is because of this principle that a dying star with a mass greater than the Chandrasekhar Limit collapses into itself, and the remaining mass transforms into a black hole.[41,42]

These principles govern all intra-cosmic black holes. Also, if the existence of a singularity in intra-cosmic black holes is accepted, before the point of singularity, the contraction of space can be relatively defined up to about the Planck length, which is the smallest meaningful unit of length in quantum physics.

However, electrons or any other constituent of matter are undefinable in the context of the Cosmic Black Hole. And since the Pauli exclusion principle and the Chandrasekhar limit are only valid with the existence of fundamental particles, these principles do not apply to the Cosmic Black Hole, which is absolute and nothing is divisible within it (Figure 17).

Figure 17: (Right image) Presence of the Pauli exclusion principle and the Chandrasekhar limit in intra-cosmic black holes. (Left image) Absence of the Pauli exclusion principle and the Chandrasekhar limit in the Cosmic Black Hole.

- **Comparison in terms of Schwarzschild Radius and Event Horizon**

As previously mentioned, general relativity predicts, at the center of every intra-cosmic black hole, there is a point with almost infinite gravity where space-time and known laws completely lose their nature. This point is called a singularity.[27] Apart from the singularity, intra-cosmic black holes have a boundary around them from which no mass or light can escape due to the very high gravitational pull. In fact, this boundary is a point of no return, known as the event horizon, and the distance between the event horizon and the singularity is called the Schwarzschild radius.[43,44] On the other hand, this radius changes based on the expansion of the universe and the motion of the black hole, influenced by its speed.

While outside the Cosmic Black Hole, i.e., after the complete contraction of the cosmos, neither space nor time will exist. Therefore, the Schwarzschild radius and consequently the internal and external event horizons, as well as other sections like the ergosphere and the quiet zone, are non-applicable and irrelevant; thus, the Cosmic Black Hole remains unchanged. In essence, this black hole is absolute (Figure 18).

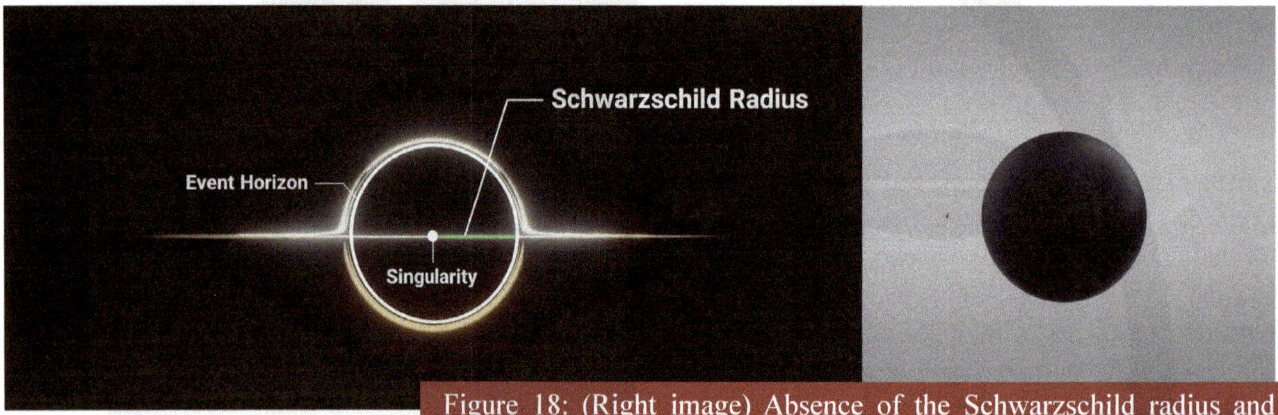

Figure 18: (Right image) Absence of the Schwarzschild radius and event horizon in the Cosmic Black Hole. (Left image) Presence of the Schwarzschild radius and event horizon in intra-cosmic black holes.

- **Comparison in terms of Velocity Relative to Space**

From the perspective of T-Consciousness Cosmology, the cosmos is in a state of rebound and everything within it is in motion with a velocity. This rebound is different from the Big Bang model's mechanism of cosmic expansion. As the cosmos increases in volume, objects within it move further apart, and the velocity of galaxies and intra-cosmic black holes also increases in the direction of this expansion. In other words, all intra-cosmic black holes are in motion and have velocities relative to each other, ultimately disintegrating and vanishing during the cosmos's final rebound moments. While the Cosmic Black Hole neither possesses motion velocity nor a counterpart to compare it to with in a relative space. Essentially, the Cosmic Black Hole is the cosmos itself in unimaginably small and infinite dimensions (Figure 19).

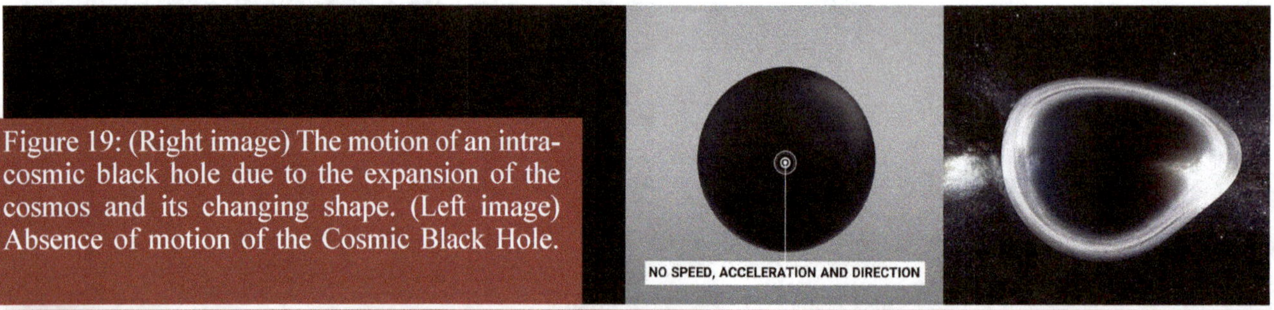

Figure 19: (Right image) The motion of an intra-cosmic black hole due to the expansion of the cosmos and its changing shape. (Left image) Absence of motion of the Cosmic Black Hole.

- **Comparison in terms of Mass and State Change**

From the viewpoint of conventional cosmology or astrophysics, based on quantum gravity theory, there is a concept known as the Hawking-Page phase transition, indicating a black hole's transition from one state to another. Scientists study the state changes of black holes in different spaces according to this concept.[45]

However, from the viewpoint of T-Consciousness Cosmology, intra-cosmic black holes are subject to motion, increasing velocity, acceleration, and direction due to their alignment with cosmic expansion; hence, this velocity does not come without consequences. That means that intra-cosmic black holes, with considerable speed, possess spin and move in synchronicity with the rebounding cosmos. Essentially, the inherent increasing velocity of cosmic rebound continuously adds to the velocity of these black holes. Consequently, any mass, including intra-cosmic black holes moving within the space mesh encounters resistance (from the mesh). This resistance leads to disintegration during space rebound (cosmic expansion). As their velocity increases along their trajectory, intra-cosmic black holes lose their ability to absorb matter, lacking the replacement mass necessary for survival. In contrast, the Cosmic Black Hole is not subject to motion, velocity or changes in state. Nothing exists beyond it to create friction and cause erosion. Therefore, it can retain absolute matter or mass within itself (Figures 20,21).

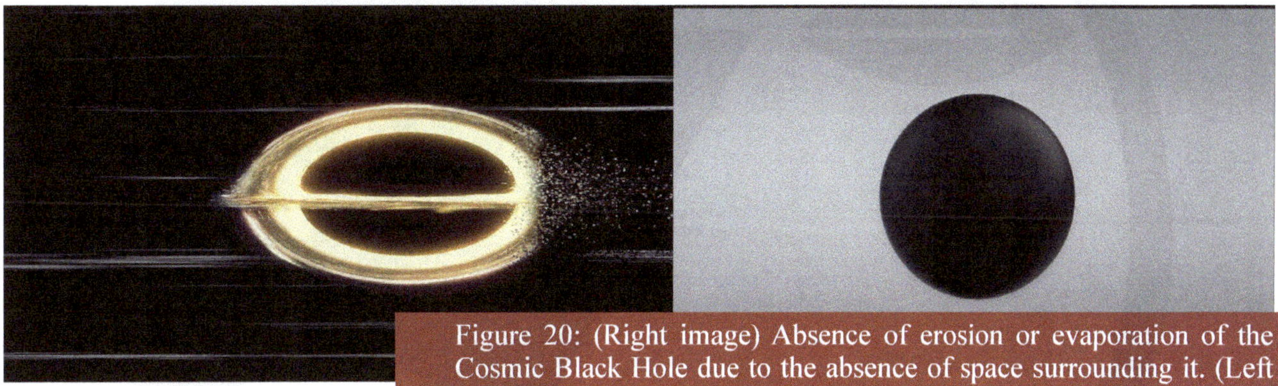

Figure 20: (Right image) Absence of erosion or evaporation of the Cosmic Black Hole due to the absence of space surrounding it. (Left image) Erosion and evaporation of intra-cosmic black holes.

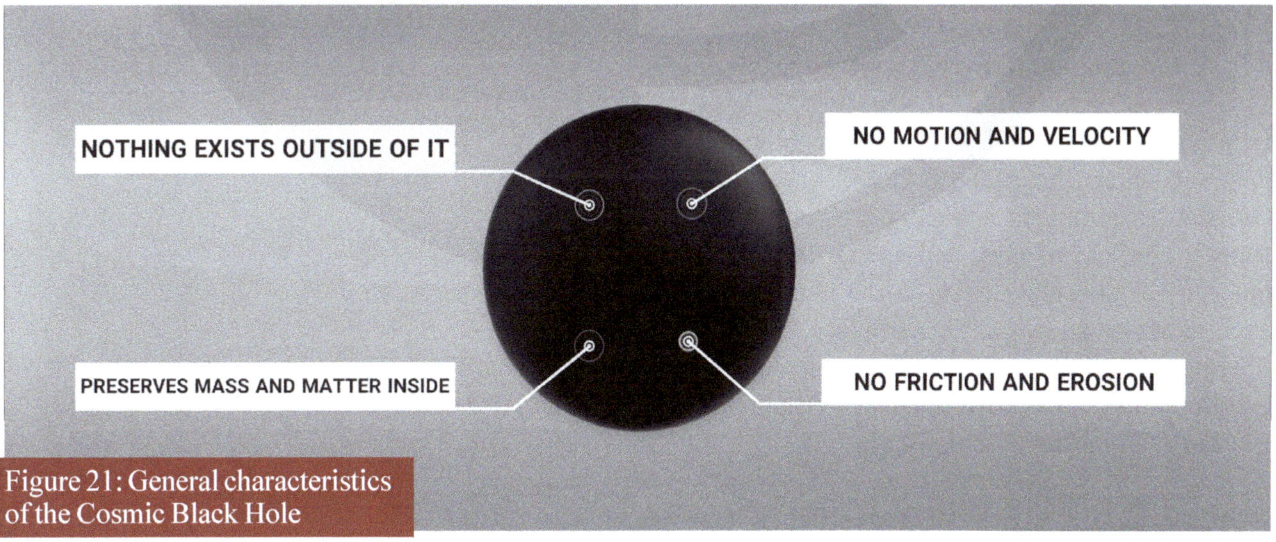

Figure 21: General characteristics of the Cosmic Black Hole

- **Comparison in terms of Temperature Exchange**

The temperature in intra-cosmic black holes is variable and surrounded by the ambient space temperature, which is equivalent to 2.7 Kelvin. In fact, there is a temperature difference between intra-cosmic black holes and the surrounding space. Therefore, the temperature exchange continues until the black hole loses its temperature and gains equilibrium with the vast cosmos. As a result, intra-cosmic black holes will one day lose their energy as they approach absolute zero temperature (Figure 22).

However, as previously mentioned, there is no space around the Cosmic Black Hole to engage in temperature and energy exchange with; hence, it cannot lose its temperature. In other words, space and time and any mass that is to emerge from this black hole in the future as components of the cosmos, are stored within it in a potential state.

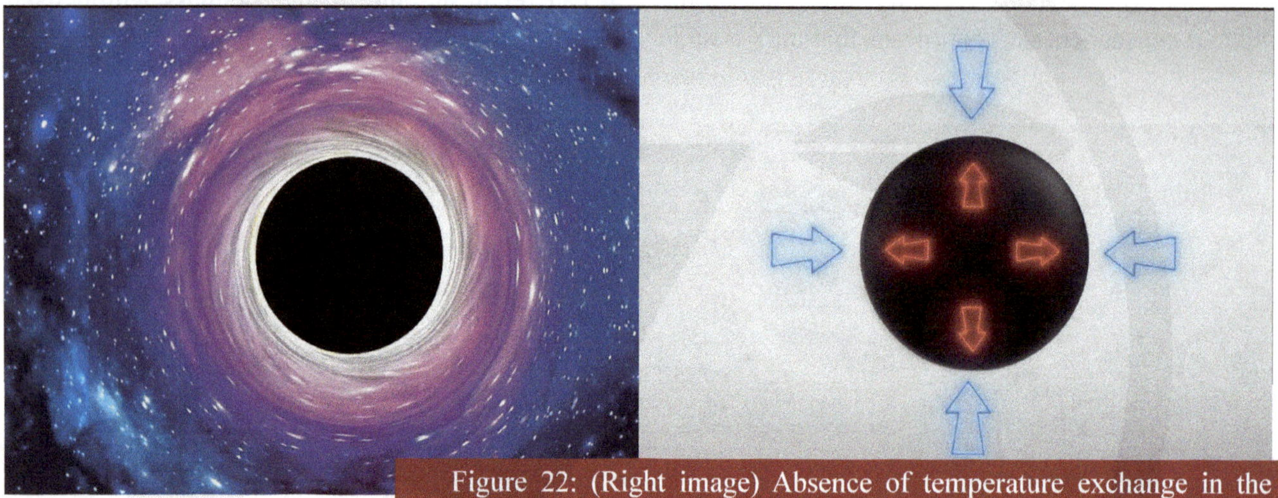

Figure 22: (Right image) Absence of temperature exchange in the Cosmic Black Hole. (Left image) Presence of temperature exchange in intra-cosmic black holes.

- **Comparison in terms of Hawking Radiation**

Intra-cosmic black holes, no matter how large, may encounter a phenomenon known as Hawking radiation. According to this theory, Hawking radiation leads to the loss of mass and energy from a black hole due to quantum effects in the vicinity of the event horizon, known as "black hole evaporation." Therefore, if this hypothesis is accepted, black holes that do not gain mass over time will evaporate and eventually vanish.[46]

However, since the Cosmic Black Hole is not in contact with any external space, the Hawking radiation hypothesis does not apply, and it cannot evaporate (Figure 23).

Journal of CosmoIntel
The First Scientific Journal in T-Consciousness Research

Figure 23: (Right image) Absence of Hawking radiation in the Cosmic Black Hole. (Left image) Presence of Hawking radiation in intra-cosmic black holes.

- **Comparison in terms of Fundamental Forces**

Scientists speculate that within a black hole, the energy might be high enough for the fundamental forces to unify with one another. However, since a complete theory of quantum gravity that unifies quantum mechanics and general relativity is not yet available, a definitive answer to this question is also currently not possible. However, it is possible that within intra-cosmic black holes, the fundamental forces, including gravitational and electromagnetic fields are all indivisible up to the hypothetical singularity point, with gravity increasing as it approaches this point.[47,48,49,50,51]

But within the Cosmic Black Hole, only gravity remains among the four fundamental forces. This near-infinite gravity crushes atomic nuclei, eliminating the other fundamental forces. As a result, the atom itself, along with any other fundamental particles, entirely lose their nature, leading to a unification of all types of matter (Figure 24).

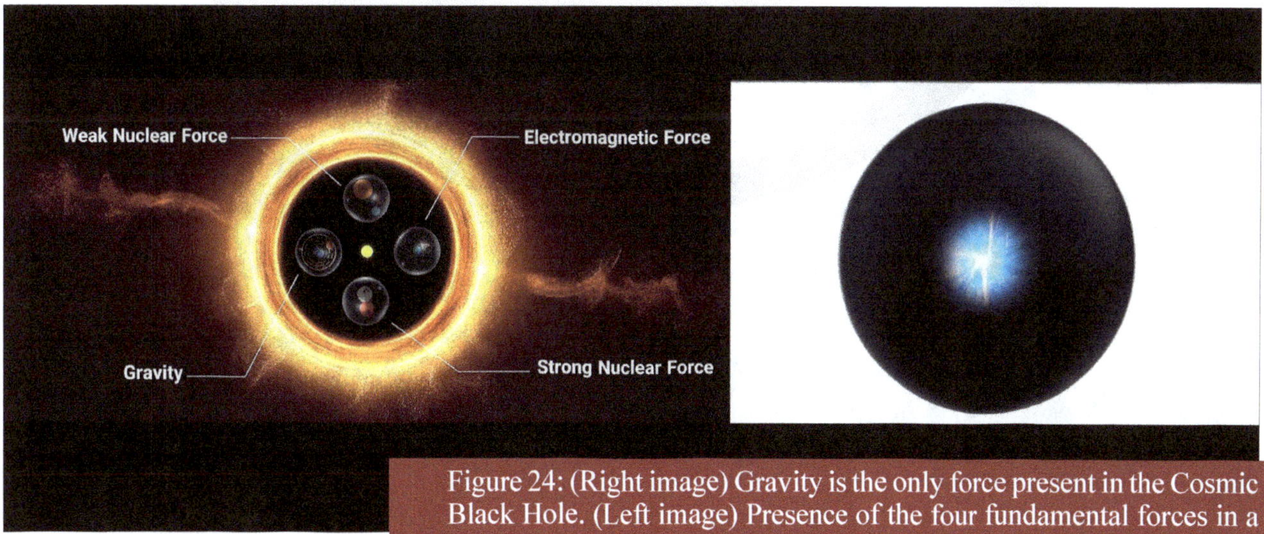

Figure 24: (Right image) Gravity is the only force present in the Cosmic Black Hole. (Left image) Presence of the four fundamental forces in a relative form in intra-cosmic black holes.

- **Comparison in terms of Time**

According to the theory of relativity, time in intra-cosmic black holes, from the perspective of an observer outside of them, is relatively dilated. [52,53,54] However, firstly, in the Cosmic Black Hole, the concept of time is introduced as an entropic force, and secondly, general relativity's definition of time, would be in its most dilated state inside the comic black hole. (Figure 25).

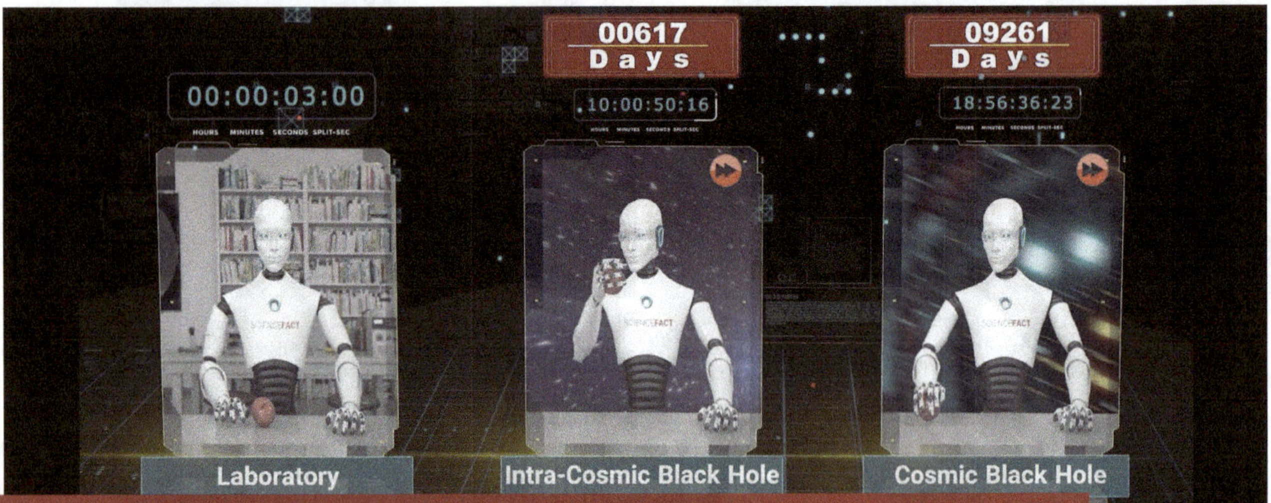

Figure 25: Passage of time in an intra-cosmic black hole and the Cosmic Black Hole.

- **Comparison in terms of Space**

Space around and inside intra-cosmic black holes is relatively compressed. In other words, space is not at its most critical and crumpled state. [55] This is while the space within the Cosmic Black Hole is in its smallest volume with ultimate compression and contraction. This compression tends toward zero but does not become zero itself, because if this was the case, the Cosmic Black Hole would vanish. On the other hand, there is no space or time around the Cosmic Black Hole to undergo contraction (Figure 26).

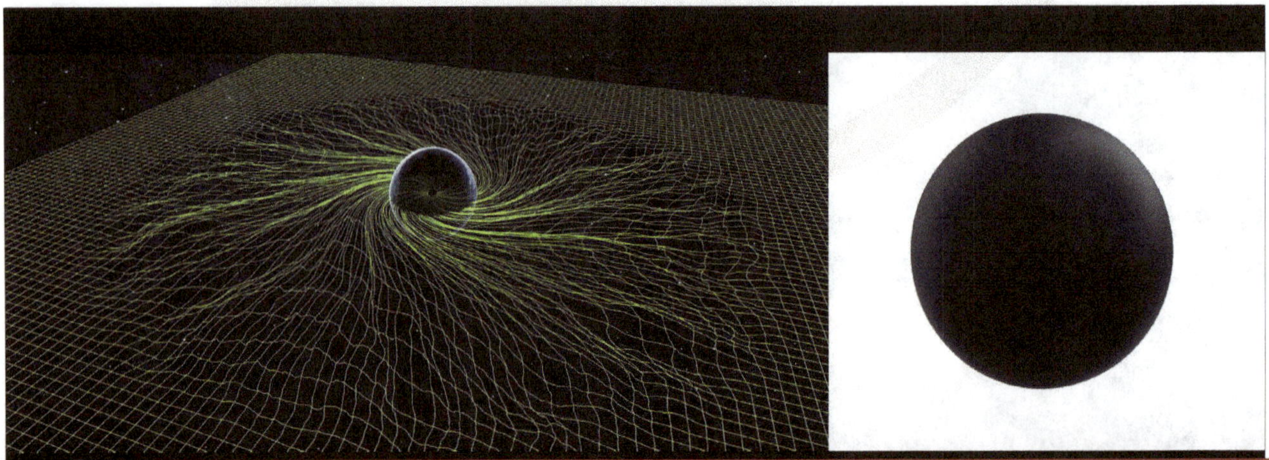

Figure 26: (Left image) Relative compression of space around and within an intra-cosmic black hole and the presence of space and time around it. (Right image) Maximum compression of space within the Cosmic Black Hole and the absence of space-time around it.

- **Comparison in terms of the Photon Sphere and Electric Charge**

Light waves spinning around powerful black holes create a halo and flow around them; this is known as the photon sphere. Some types of intra-cosmic black holes have a photon sphere and electric charge. [56,57,58,59] While the Cosmic Black Hole, due to the absence of space and also the lack of energy and fundamental particles in its vicinity, cannot have a photon sphere and electric charge (Figure 27).

Figure 27: (Left image) Lack of a photon sphere and electric charge in the Cosmic Black Hole. (Right image) Presence of a photon sphere and electric charge in intra-cosmic black holes.

- **Comparison in terms of mergeability**

Intra-cosmic black holes can merge with each other, and when they come close, they transform into a larger black hole with greater mass and stronger gravity through the process of merging. [60,61,62] However, the Cosmic Black Hole is completely unique and solitary, with no other mass outside of it that it could potentially merge with (Figure 28).

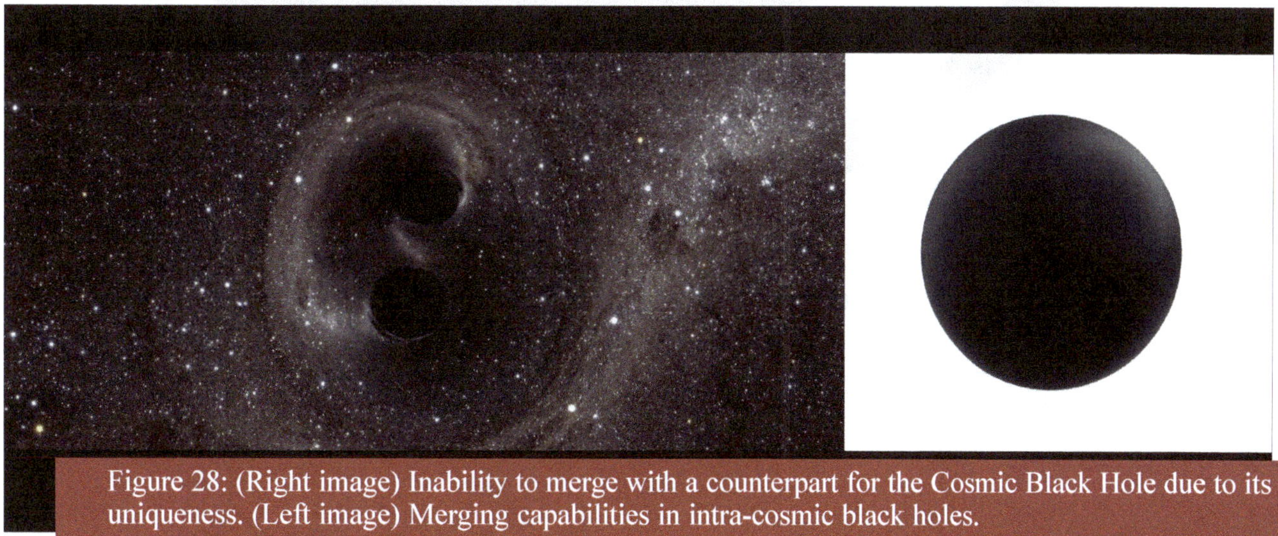

Figure 28: (Right image) Inability to merge with a counterpart for the Cosmic Black Hole due to its uniqueness. (Left image) Merging capabilities in intra-cosmic black holes.

References

[1] Jones, A. Z. (2019, January 29). *Description & Origins of Inflation Theory*. ThoughtCo. www.thoughtco.com/what-is-inflation-theory-2698852

[2] Guth, A. H. (2004). Inflation. In Freedman, W. L. (Eds.), *Measuring and Modeling the Universe* (pp. 31-51). Cambridge: Cambridge University Press.

[3] Gleiser, M. (2023, May 24). *Is the Universe a quantum fluctuation?* Big Think. https://bigthink.com/13-8/universe-quantum-fluctuation/

[4] Harvard & Smithsonian, Center for Astrophysics. (n.d.). *What happened in the early universe?* https://pweb.cfa.harvard.edu/big-questions/what-happened-early-universe

[5] Maziashvili, M. (2006). *Quantum Fluctuations of Space-Time. arXiv*, https://arxiv.org/abs/hep-ph/0605146

[6] Wolchover, N. (2019, June 6). *Physicists Debate Hawking's Idea That the Universe Had No Beginning.* Quantamagazine. https://www.quantamagazine.org/physicists-debate-hawkings-idea-that-the-universe-had-no-beginning-20190606/

[7] Page, D. N. (2006). *Susskind's Challenge to the Hartle-Hawking No-Boundary Proposal and Possible Resolutions. arXiv*, https://arxiv.org/abs/hep-th/0610199

[8] Hawking, S. W. (1993). The no-boundary proposal and the arrow of time. *Vistas in Astronomy, 37*, 559-568.

[9] Feldbrugge, J. L., Lehners, J. L., & Turok, N. (2017). No rescue for the no boundary proposal: Pointers to the future of quantum cosmology. *Physical Review D, 97*, 023509.

[10] Britannica, T. Editors of Encyclopaedia (2012, January 9). *Friedmann universe. Encyclopedia Britannica.* https://www.britannica.com/science/Friedmann-universe

[11] Schwarzschild, K. (1999). On the Gravitational Field of a Sphere of Incompressible Fluid according to Einstein's Theory (S. Antoci., Trans.). Sitzungsberichte der Königlich Preussischen Akademie der Wissenschaften [Berlin]. (Original work published in 1916). https://arxiv.org/pdf/physics/9912033.pdf

[12] NASA. (n.d.). *Black Holes*. NASA Science. Retrieved from https://science.nasa.gov/universe/black-holes/

[13] Britannica, T. Editors of Encyclopaedia (2023, December 22). *Event horizon. Encyclopedia Britannica.* https://www.britannica.com/topic/event-horizon-black-hole

[14] Lea, R., & Choi, C. Q. (2023, March 4). *What is a black hole event horizon (and what happens there)?* Space.com. https://www.space.com/black-holes-event-horizon-explained.html

[15] Schutz, B. (2003). *Gravity from the Ground Up: An Introductory Guide to Gravity and General Relativity*. Cambridge: Cambridge University Press.

[16] Davies, P. C. W. (1978). Thermodynamics of Black Holes. *Reports on Progress in Physics*, *41*(8), 1313–1355.

[17] NASA Science Editorial Team. (2019, September 23). *10 Questions You Might Have About Black Holes*. NASA. https://science.nasa.gov/universe/10-questions-you-might-have-about-black-holes

[18] Nakazato, K. i., Sumiyoshi, K., Suzuki, H., & Yamada, S. (2008). Oscillation and future detection of failed supernova neutrinos from a black-hole-forming collapse. *Physical Review D, 78*(8), 083014.

[19] Kodama, H. (2004). *Perturbative Uniqueness of Black Holes near the Static Limit in All Dimensions*. *Prog.Theor.Phys. 112* (2004) 249-274. *arXiv*, https://arxiv.org/abs/hep-th/0403239

[20] Freudenrich, C. (2023, September 8). *How Black Holes Work*. Howstuffworks. https://science.howstuffworks.com/dictionary/astronomy-terms/black-hole.htm

[21] Oxford Reference (n.d). Static limit. In *A Dictionary of Astronomy* [Online]. Retrieved from https://www.oxfordreference.com/display/10.1093/oi/authority.20110803100529279

[22] Griest, K. (Spring 2014). *Physics 161: Black Holes. Section 14.4: The Ergosphere* [Course Book]. University of California. Retrieved from https://courses.physics.ucsd.edu/2014/Spring/physics161/book.pdf

[23] NASA. (n.d.). *Anatomy of a Black Hole*. https://science.nasa.gov/universe/black-holes/anatomy/

[24] NASA. (2019, September 25). *NASA Visualization Shows a Black Hole's Warped World*. https://www.nasa.gov/universe/nasa-visualization-shows-a-black-holes-warped-world/

[25] Lea, R. (2023, June 28). *Distortions in space-time could put Einstein's theory of relativity to the ultimate test*. *LiveScience*. https://www.livescience.com/physics-mathematics/quantum-physics/distortions-in-space-time-could-put-einsteins-theory-of-relativity-to-the-ultimate-test

[26] Sobral-Blanco, D., & Bonvin, C. (2023). Measuring the distortion of time with relativistic effects in large-scale structure. *MNRAS Letters 519*, L39 (2023). *arXiv*, https://arxiv.org/abs/2205.02567

[27] Baird, C. S. (2013, September 13). *Does every black hole contain a singularity?* Science Questions with Surprising Answers. https://www.wtamu.edu/~cbaird/sq/2013/09/13/does-every-black-hole-contain-a-singularity/

[28] Turner, C. (2019, April 14). *Astronomers Reveal First Direct Visual Evidence of a Supermassive Black Hole*. SciTechDaily. https://scitechdaily.com/astronomers-reveal-first-direct-visual-evidence-of-a-supermassive-black-hole/

[29] Lu, H., & Lyu, H. (2019). On the Size of a Black Hole: The Schwarzschild is the Biggest. *Phys. Rev. D 101*, 044059. *arXiv*, https://arxiv.org/abs/1911.02019

[30] Bel, Ll. (2007). Uber das Gravitationsfeld eines Massenpunktes nach der Einstenschen Theorie. *arXiv*, https://arxiv.org/abs/0709.2257

[31] Wang, X., Li, R., & Wang, J. (2021). Islands and Page curves of Reissner-Nordström black holes. J. *High Energ. Phys*. 2021, *103*. *arXiv*, https://doi.org/10.48550/arXiv.2101.06867

[32] Teukolsky, S. A. (2015). The Kerr Metric. *Classical and Quantum Gravity*, *32*(12), 124006. https://doi.org/10.1088/0264-9381/32/12/124006

[33] Newman, E. T., & Janis, A. I. (1965). Note on the Kerr Spinning-Particle Metric. *Journal of Mathematical Physics*, *6*(6), 915–917. file:///C:/Users/64204/Downloads/J_Math_Phys_Newman_1_copy.pdf

[34] Drake, S. P., & Szekeres, P. (1998). An explanation of the Newman-Janis Algorithm. *Gen.Rel. Grav.32*:445-458,2000. *arXiv*, https://doi.org/10.48550/arXiv.gr-qc/9807001

[35] Adamo, T., & Newman, E. T. (2014). The Kerr-Newman metric: A Review. General Relativity and Quantum Cosmology. *Scholarpedia 9*: 31791, 2014. *arXiv*, https://doi.org/10.48550/arXiv.1410.6626

[36] Kramer, H. J. (2019, April 12). *Black Hole Image 2019*. eoPortal, Astronomy and Telescopes. https://www.eoportal.org/other-space-activities/black-hole#before-and-after

[37] May, A. (2021, August 25). *8 ways we know that black holes really do exist*. Live Science. https://www.livescience.com/how-we-know-black-holes-exist.html

[38] Clegg, B. (2020, July 4). *What is a black hole and how did we discover them?* BBC Science Focus. https://www.sciencefocus.com/space/black-holes

[39] Grossman, D. (2019, April 10). *How They Got the Black Hole Picture That Changed Science*. Popular Mechanics. https://www.popularmechanics.com/space/deep-space/a27099934/eht-black-hole-picture

[40] Kaplan, I. G. (2019). Pauli Exclusion Principle and its theoretical foundation. *arXiv*, https://arxiv.org/abs/1902.00499

[41] Bethe, H. A., Brown, G. E, & Lee, C. H. (2003). How A Supernova Explodes. In *Formation And Evolution of Black Holes in the Galaxy: Selected Papers with Commentary* (pp. 51-61). World Scientific Publishing Company.

[42] Harvard & Smithsonian Center for Astrophysics. (n.d.). *Black Holes*. https://www.cfa.harvard.edu/research/topic/black-holes

[43] Lü, H., & Lyu, H. D. (2020). Schwarzschild black holes have the largest size. *Physical Review D*, *101*(044059). https://doi.org/10.1103/PhysRevD.101.044059

[44] Britannica, T. Editors of Encyclopaedia (2024, March 20). *Schwarzschild radius. Encyclopedia Britannica*. https://www.britannica.com/science/Schwarzschild-radius

[45] Su, B. Y., Wang, Y. Y., & Li, N. (2019). The Hawking-Page phase transitions in the extended phase space in the Gauss-Bonnet gravity. *arXiv*, https://doi.org/10.48550/arXiv.1905.07155

[46] Ghosh, A. (2019). Hawking Radiation – Revisited. *arXiv*, https://arxiv.org/pdf/1901.10069.pdf
Gold (https://physics.stackexchange.com/users/21146/gold), Are the fundamental forces unified in a black hole?, URL (version: 2017-03-23): https://physics.stackexchange.com/q/316629

[47] Husain, V., Kelly, J. G., Santacruz, R., & Wilson-Ewing, E. (2022). Quantum Gravity of Dust Collapse: Shock Waves from Black Holes. *Physical Review Letters*, *128*(121301). https://doi.org/10.1103/PhysRevLett.128.121301

[48] Pound, A., & Wardell, B. (2021). Black hole perturbation theory and gravitational self-force. *arXiv*, https://doi.org/10.48550/arXiv.2101.04592

[49] Ross, G. (2003). *Grand Unified Theories*. Boulder, Colorado: Westview Press.

[50] Georgi, H., & Glashow, S. L. (1974). Unity of All Elementary-Particle Forces. *Physical Review Letters*, *32*, 438.

[51] Einstein, A. (2011). *Relativity – The Special and General Theory* (Originally published in 1916). Read Books Ltd.

[52] Profound Physics. (n.d.). *Why Time Slows Down Near a Black Hole*. https://profoundphysics.com/why-time-slows-down-near-a-black-hole/

[53] Annika Peterson (https://physics.stackexchange.com/users/4179/annika-peterson), How exactly does time slow down near a black hole?, URL (version: 2014-03-26): https://physics.stackexchange.com/q/25759

[54] LibreTexts. (n.d.). *Spacetime Near Black Holes*. In Coble, K., et al. (Eds.), *Big Ideas in Cosmology*. Retrieved from https://phys.libretexts.org/Bookshelves/Astronomy__Cosmology/Big_Ideas_in_Cosmology_(Coble_et_al.)/11%3A_Black_Holes/11.02%3A_Spacetime_Near_Black_Holes

[55] Wang, H.-T., Li, P.-C., Jiang, J.-L., Yuan, G.-W., Hu, Y.-M., & Fan, Y.-Z. (2021). Constrains on the electric charges of the binary black holes with GWTC-1 events. *The European Physical Journal C*, *81*(8), 769.

[56] Ted Bunn (https://physics.stackexchange.com/users/1241/ted-bunn), *Detection of the Electric Charge of a Black Hole: How can an electromagnetic field escape the event horizon of a Reissner-Nordström black hole?*, URL (version: 2011-07-12): https://physics.stackexchange.com/q/12171

[57] Zaumen, W. T. (1974). Upper bound on the electric charge of a black hole. *Nature, 247*(5442), 530-531. doi:10.1038/247530a0

[58] Zajaček, M., & Tursunov, A. (2019). Electric charge of black holes: Is it really always negligible? *arXiv*,https://arxiv.org/abs/1904.04654

[59] Lea, R. (2022, May 31). *A black hole formed by a lopsided merger may have gone rogue*. Space.comhttps://www.space.com/black-hole-escaping-galaxy-from-collision

[60] Varma, V., Biscoveanu, S., Islam, T., Shaik, F. H., Haster, C.-J., Isi, M., . . . Vitale, S. (2022). Evidence of Large Recoil Velocity from a Black Hole Merger Signal. *Physical Review Letters, 128*(19), 191102. doi:10.1103/PhysRevLett.128.191102

[61] Raccanelli, A., Kovetz, E. D., Bird, S., Cholis, I., & Muñoz, J. B. (2016). Determining the progenitors of merging black-hole binaries. *Physical Review D, 94*(2), 023516. doi:10.1103/PhysRevD.94.023516

[62] Carr, B. (Ed.). (2010). *Universe or Multiverse?* Cambridge: Cambridge University Press.

[63] Mersini-Houghton, L. (2023). *Before the Big Bang: The Origin of the Universe from the Multiverse*. Mariner Books.

[64] O'Raifeartaigh, C., McCann, B., Nahm, W., & Mitton, S. (2014). Einstein's steady-state model of the universe. *arXiv*, https://arxiv.org/abs/1402.0132

[65] Becker, K., Becker, M., & Schwarz, J. H. (2006). *String Theory and M-Theory: A Modern Introduction.* Cambridge: Cambridge University Press.

[66] Bojowald, B. (2012). Loop quantum gravity and cosmology. In Murugan, J., Weltman, A. & George F. R. Ellis (Eds.), *Foundations of Space and Time: Reflections on Quantum Gravity* (pp. 211–256). Cambridge: Cambridge University Press.